Life in the Digital Bubble

Life in the Digital Bubble

A Digital Visionary's perspective on the past, present, and future of digital technology—offering a blueprint for individuals, families, and societies to not just survive but thrive in the next three decades

TAMER BADAWY

Dedication

To my parents, whose values and wisdom continue to guide me.
To my wife and children, whose love and questions keep me grounded in a world of screens.
And to the curious minds seeking clarity in the chaos of the digital age.

Life in the Digital Bubble

ISBN: 978-2-9701936-0-9 (Hardcover)

Published independently by Tamer Badawy

Printed and distributed by IngramSpark

www.tamerbadawy.com

First edition, September 2025

Table of Contents

Introduction

The loud buzz of an alarm clock broke through my sleep. Groaning, I reached for my smartphone, but my hand met wood instead of smooth glass. Confused, I opened my eyes to see an old, clunky alarm clock glowing red at 5:30 AM, the same one I used to wake up to during my university days.

What was going on?

I bolted upright, heart pounding. The room was unfamiliar. No smart blinds gliding open to reveal the neighborhood. No soft, ambient lighting. Just beige curtains and a single yellow bulb overhead. I rubbed my face, trying to make sense of it. Was I on vacation in some retro-themed hotel?

No AI assistant. No holographic calendar. Just silence.

I swung my legs off the bed, my feet landing on a carpet instead of the warm wooden floors I was used to. A shiver ran through me.

Then I saw it: a rotary phone on the bedside table, its coiled cord neatly arranged.

"You've got to be kidding me. What is this?"

I stumbled to the bathroom and stared into the mirror, half-expecting a glitch or distortion, but there was nothing. It was my face, just like yesterday. What caught my eye, though, was my T-shirt with a faded Blockbuster Video logo.

I took a deep breath. Maybe this was an elaborate prank. Maybe my friend Husein had gone overboard with a VR nostalgia simulation.

I turned on the faucet. Cold water rushed out. I splashed my face, hoping to snap out of this madness, but it was all too real.

As I dried my face, something shiny in the medicine cabinet caught my eye. I opened it and found a pager.

A pager?

I picked it up, turning it over in my hands. I hadn't seen one of these since the mid-90s.

The realization struck me hard. Somehow, I had gone to bed in 2025 and woken up thirty years in the past.

Dazed, I walked back to the bedroom and opened the closet. Flannel shirts, baggy jeans, and oversized sweaters stared back at me. I grabbed a plain navy polo and khakis.

Dressed, I headed to the kitchen. The scratched linoleum floor and the loud hum of the fridge greeted me. I opened the door, expecting a smart interface to tell me I was out of oat milk or suggest meal plans. Instead, I found dimly lit, empty shelves.

I stood there, frustrated. No voice-activated grocery lists or energy-efficient controls, just an old fridge with a stubborn door.

The coffee maker was ancient compared to the sleek one I was used to, the kind that could whip up a cold foam cappuccino. I filled it manually and waited as it gurgled and brewed weak coffee.

No voice commands. No instant customization. Just a slow, clunky process.

As I sipped the bitter brew, I instinctively glanced at the wall, expecting a display showing my commute or the weather. Nothing. No traffic updates. No smart assistant to tell me when to leave for work.

I reached for my non-existent smartwatch, realizing with a sinking feeling that I was completely on my own.

I leaned back in my chair, thinking hard. In 2025, I could summon any information with a touch. I had a self-driving car, AI assistants managing my schedule, and smart home systems handling my mornings.

Now? I was staring at a rotary phone and an empty fridge.

It was slow. Frustrating. But I had no choice.

I squared my shoulders. I had to figure this out, but first I needed to survive the day.

Suddenly, the world shimmered. Everything blurred and flickered, like a glitching computer screen. The rotary phone, the pager, the faded wallpaper—it all dissolved.

I gasped as the sensation of falling pulled me into darkness.

BZZZT! BZZZT!

My iPhone buzzed on the nightstand. I shot upright, drenched in sweat, my heart racing.

The sleek glass screen glowed: 5:30 AM. The smart blinds slid open, revealing the early morning sky.

I blinked, still shaken. It had all been a dream, a dream within a dream.

As I silenced the alarm, I let out a shaky laugh. The future was right here in my hands.

Despite the panic I had felt in the dream, it left me with a strange sense of nostalgia. It was nostalgia for the past. I couldn't help but grab my favorite morning cappuccino and head to the office room, where I pulled out an old photo album. The pages were filled with faded memories: black-and-white family photos, handwritten letters, and postcards from friends who had moved away. I paused at a picture of me and some friends laughing on a balcony. Back then, if I wanted to talk to anyone in that photo, I had to walk to their house or write them a letter. Now, everything is so instant.

The thought lingered as I looked down at my phone, buzzing with notifications. Emails, messages, and social media updates filled the screen. In my hand was a device that could connect me to anyone, anywhere, in seconds. But at what cost?

Over the past few decades, something profound has happened. The way we communicate, connect, and even think has been transformed by a force so prevailing that we often don't even notice it. It is not just the internet or smartphones; it is something deeper. I call it the *Digital Bubble*, an invisible, intelligent network that surrounds each of us.

This bubble did not exist a few decades ago, but today it is as essential as the air we breathe. It determines how we work, relate to others, learn, and navigate the world. And it is growing smarter every day.

As I sat there, sipping my coffee and gazing out the window at the ice-capped Alps, my mind raced. The changes between 1995 and 2025—the rise of the internet, smartphones, artificial intelligence, and the internet of things—have been staggering. But what struck me most was realizing that these changes, as monumental as they seem, are just the beginning.

The pace of technological advancement is accelerating exponentially. What took 30 years to achieve, from the analog world of 1995 to the digital bubble of 2025, might take only 5 or 10 years to surpass. By 2055, the world could be unrecognizable.

My heart started to race again as a wave of anxiety washed over me. What would the world look like in 2055? How could I possibly keep up with such relentless change? More importantly, how could I not just survive but truly thrive in this new reality? The questions flooded my mind faster than I could process them. Would we still use smartphones, or would brain-computer interfaces (BCIs) make thoughts our primary mode of communication? Would AI and automation replace most jobs, forcing us to redefine purpose and fulfillment?

Then there was the question of relationships. Would VR and AI companions replace human connection, or would they somehow deepen it? A shiver ran down my spine as I considered privacy and security. As the digital bubble thickened, would we find ways to protect our autonomy, or would we become mere data points in an all-knowing system?

My thoughts spiraled further. How would society adapt? Would technology bring prosperity to all, or would the gap between the privileged and the forgotten widen even more? The future loomed ahead, both dazzling and daunting, and I realized these were not just abstract questions. They were the defining issues of our time.

The more I thought about it, the more overwhelmed I felt. The changes we have experienced so far are just a glimpse of what is to come. By 2055, the digital bubble could be so advanced and integrated into our lives that it is difficult to imagine what "normal" will even look like.

As I tried to process these thoughts, my eyes drifted back to the Alps outside my window. Their timeless beauty stood in stark contrast to the whirlwind of change swirling in my mind. Then, amidst the anxiety, a new question emerged. What can I do to prepare for this future, not just for myself but for everyone I care about?

The answer came to me slowly but clearly. This is not just about coping with change; it is about embracing it. The digital bubble is not going away. It is only going to grow thicker by the day. That can be terrifying, but it is also an opportunity.

A sudden clarity washed over me. Thriving in this new world is not just about keeping up with technology. It is about understanding it, shaping it, and using it to enhance our lives rather than letting it control us. The key is not blind adaptation but intentional mastery.

I needed to ask the right questions. How could I stay informed and educated about emerging technologies without feeling overwhelmed? How could I use these tools not just for convenience but to improve my work, my relationships, and my well-being? Perhaps the most important question of all was how I could contribute to shaping a future that is ethical, just, and sustainable. These were not idle thoughts. They were the foundation of a new way of living, a roadmap to navigating the unknown with purpose and agency.

At that moment, my perspective shifted. Instead of feeling overwhelmed, I felt a sense of purpose. The digital bubble is not just something that is happening to us. It is something we can actively shape.

As I finished my coffee and prepared to head to the office, I made a decision. I would not just sit back and let the future happen. I would engage with it, learn from it, and strive to make the most of it. The changes between 1995 and 2025 were just the beginning. The next 30 years will be even more transformative, and I want to be ready.

But this is not just about me. I felt a deep responsibility to share what I had learned, to tell everyone I knew what to expect in the coming decades. We all need to be prepared. We all must act.

The world is changing faster than ever, and we cannot afford to be passive observers. We must rethink how we live, work, and raise the next generation. How do we prepare our children for a future where the digital bubble is even more pervasive and powerful? How do we teach them to balance their lives, to harness technology without being consumed by it?

Both the questions and their answers inspired me to write this book. It is more than a reflection on the past or a prediction of the future. It is a call to action. We have the power to shape the digital bubble, ensuring it serves humanity rather than controls it. Many harmful forces will attempt to use it for their own gain, and we must be ready to counter them. We must act now, with intention and purpose.

One thing is certain. The future is not just something we are heading toward. It is something we are building together. And I, for one, want to help shape it for the better.

Chapter 1
Before the Digital Bubble

Analog Roots Tracing the History of Human Communication

It is all about communication!

We must step back, far back, if we are to fully comprehend the digital bubble that now governs the majority of our lives. Direct, immediate communication has existed from the start of humanity, with no intermediaries or screens. Imagine this: A hunter, breathless and exhausted, returns to his clan after narrowly escaping a deadly beast. He gestures wildly to describe the creature's movements, his hands mimicking the beast's furious charge. His voice rises and falls, carrying the emotions of danger, fear, and triumph. The tribe watches with wide eyes, completely captivated by the storyteller. This form of communication was raw, personal, and powerful. There were no channels or gadgets designed to skew the meaning. The hunter made every statement and action come to life. This was communication in its most primal form,[1] a human-to-human connection that relied solely on the body and mind to convey meaning.

As time passed, and as communities grew larger and more complex, so did the need for more sophisticated methods of communication. The invention of writing was a monumental turning point in this evolution. Suddenly, communication was no longer confined to the present moment; messages could be recorded, saved, and passed on across time and space. Picture an ancient scribe hunched over a clay tablet, carefully etching symbols into its surface, preserving the results of a harvest or a tale of a great battle. Early civilizations in Mesopotamia, Egypt, and China began to use written symbols to convey ideas, laws, and instructions.[2] It was a revolutionary step, allowing ideas to be captured and shared without the limitations of time or distance.

The invention of the printing press in the 15th century, by Johannes Gutenberg, marked yet another revolution in the way people communicated.[3] Before the press, books were rare and expensive, often confined to the hands of a few wealthy scholars. Gutenberg's press changed all of that. Suddenly, books could be mass-produced, making knowledge available to the masses. Imagine a bustling 16th-century marketplace filled with merchants and vendors eagerly selling freshly printed pamphlets. These pamphlets, filled with revolutionary ideas and news, spread rapidly throughout towns and cities, marking the beginning of a new era of information sharing. The world was forever changed as the written word could now travel faster and farther than ever before.

Then, in the 19th century, came a breakthrough that would change communication forever: telecommunications. The invention of the telegraph allowed for messages to be transmitted across great distances almost instantaneously.[4] Imagine a family sitting anxiously in a telegraph office, waiting to hear news from a loved one who lived far away. The tapping of Morse code filled the room as they eagerly awaited a message. Soon, the telephone would emerge, making it possible for people to hear each other's voices across vast distances. For the first time, communication was no longer constrained by geography.

Then, as the 20th century progressed, the invention of the radio brought an entirely new dimension to communication. For the first time, people could hear voices, music, and stories broadcast into their homes. Families gathered around radios to listen to the news, follow their favorite programs, and hear live concerts. It was a communal experience, one that connected people in ways they had never imagined before. My parents always told me about the streets of Cairo, my hometown, filled with families gathering around radios, eagerly awaiting the start of an Om Kulthum concert.[5] Om Kulthum, the legendary Egyptian singer, held the hearts of millions in the Arab world. Her voice transcended generations, and her monthly radio concerts united people from all backgrounds.

The invention of television introduced a visual dimension to communication, transforming it from just sound into a full sensory experience. For the first time, people could not only hear the events unfolding in the world, but they could also see them. The 1980 Moscow Olympics, for example, became a defining moment for millions of people around the world, who gathered in front of their television sets to watch the opening ceremony.

But then, in the late 20th century, came the internet—and everything changed. The digital revolution was upon us, and the world would never be the same again. Information could now be shared instantly, across vast distances, at the click of a button.

The world was becoming smaller, and communication was no longer limited to physical objects or the constraints of time.

1995 When the World Was About to Change

It didn't happen overnight, but when it did, it completely transformed everything. The invention of the internet in the late twentieth century signaled the beginning of a new era[6] in which information could be instantly shared across vast distances. But, before we get into the specifics of this digital revolution, let's take a moment to consider how we lived in 1995. Although the world was on the verge of transformation, communication, entertainment, and daily life continued to move at a slower, more deliberate pace. The technology that would soon shape our lives was still in its infancy, and many of the conveniences we take for granted today were mere concepts.

In 1995, communication was slow, deliberate, and frequently deeply personal. If you wanted to communicate with someone, your options were limited but meaningful. Most households relied on landline phones, which were frequently shared by entire families or even buildings. Long-distance calls were expensive, so conversations were brief and to the point. For those separated by distance, handwritten letters and postcards were the primary means of communication, with each message conveying a sense of effort and intention. Another way to bridge the gap was to record a personal message and mail it to loved ones, as long-distance calls were extremely costly. If you wanted to see someone, you had to meet them in person because there were no video calls or instant messaging to bridge the gap. Every interaction, whether by phone, letter, or in person, felt genuine, with the presence and authenticity that defined the era.

I remember making such tapes for my uncle, who lived across the ocean. It was more than just sending a message; it became an event in our large family home. The entire family would gather around, knowing that this audiotape would travel over 10,000 kilometers to deliver our voices to him. We would choose our words carefully, ensuring that the tape felt personal and meaningful. When the tape finally arrived, hearing his voice on the other end and knowing it had traveled across oceans made the distance feel a little less daunting. It was a unique ritual that filled the house with excitement and connection in ways that felt completely genuine.

The workplace was vastly different from today, with a slower pace and an emphasis on physical processes rather than digital tools. Most people worked in offices or factories where technology played little role. Paper-based systems were common, with documents printed, filed, and stored in cabinets, and while email existed, it was not widely used. Employees typically adhered to fixed working hours, clocking in for fixed shifts with little flexibility, and the concept of remote work was virtually unknown. Access to information was also limited; if you needed answers, you went to libraries to look up books and encyclopedias, or you relied on the knowledge of colleagues or friends, because there was no Google or Wikipedia to provide instant answers. The workplace in 1995 was one of tangible files, strict schedules, and face-to-face collaboration.

In terms of entertainment, 1995 was a year full of shared experiences that brought people together. Families and friends would gather around the television, eagerly anticipating their favorite shows, or go to the movies to see the latest releases. Television and radio were the primary sources of entertainment and news. With only a few channels, shows were scheduled at specific times, so if you missed one, you could either wait for a rerun or find someone who could recap it for you over coffee. When it came to music, people were passionate about their CD collections or cassette tapes, and sharing a song entailed physically lending a copy to a friend. The news, too, relied heavily on traditional sources such as printed newspapers, as well as television and radio broadcasts. Every morning, people would pick up a newspaper, read the headlines, and compile the day's most important stories. Flipping through pages was a ritual, whether reading at the breakfast table or discussing the day's news over coffee.

Socializing was always done in person, whether at home, in a coffee shop, in the park, or at local events, where people spent time together face to face, making connections that felt immediate and personal. This sense of community and connection, while limited by the technology of the time, resulted in moments of intimacy and shared experience that we now take for granted. Real-world interactions were at the center of social life. Community events such as local festivals, fairs, and religious gatherings brought people together, instilling a sense of belonging and community. Friendships were maintained through regular meetings and phone calls. Personal moments were truly private, with no pressure to share every detail with the world. In this analog world, relationships were based on trust and time, and the bonds felt deeply ingrained.

Despite its limitations, the analog world had its advantages. Communication, while slower, was frequently more meaningful. People were not distracted by smartphones or the never-ending scroll of social media; when they interacted, they were completely present. Every conversation had an air of quiet intimacy. Personal lives remained private, with no digital footprint to track every movement. It was a time when individuals had complete control over what they shared and with whom.

Education in 1995 was a practical, hands-on experience. Classrooms were packed with students attending lessons in person, with teachers writing on chalkboards and students taking notes by hand. Research was an adventure in libraries, where students spent hours flipping through card catalogs looking for books to help them understand. Group projects required in-person meetings, which were often held at someone's home or in the library, where collaboration occurred face-to-face. There were no online tools like Google Docs or Zoom, only the quiet hum of teamwork and the effort required to physically meet and collaborate.

Shopping in 1995 was a hands-on, in-person experience, with people visiting stores to buy everything from groceries to clothing. Malls, though fewer in number, were vibrant social hubs at the time, teeming with friends and families who spent hours browsing the shops, grabbing a bite to eat, or catching up while wandering between stores. The atmosphere was lively, with the sounds of crowds and shopping carts filling the air. The product selection was more limited than today's vast online marketplaces, and if you couldn't find exactly what you were looking for, you'd frequently have to visit multiple stores, relying on shopkeepers' expertise to guide your search. Even as credit cards began to appear in more wallets, offering a glimpse into the future of transactions, paying with cash remained the norm. This tactile, in-person shopping experience helped us connect more deeply with the items we bought and the people we met along the way. It was more than just buying things; it was about the journey, the discovery, and the moments of connection that came with each shopping trip.

Travel in 1995 was characterized by careful planning, effort, and a spirit of adventure. Because online booking platforms were still in their infancy, it was common to rely on a travel agent to help arrange flights, hotels, and tours when planning a trip. Planning your itinerary entailed spending countless hours researching destinations, reading travel brochures, and consulting guidebooks. Lonely Planet was the go-to resource for travelers looking for tips on everything from must-see attractions to local customs and hidden gems. Having the most recent edition of these guidebooks was critical, especially if you found yourself stranded in the middle of nowhere and needed the most up-to-date information to navigate unfamiliar territory.

Paper maps were essential for navigating new destinations. GPS devices were rare, expensive, and difficult to obtain, so relying on your own sense of direction or asking locals for assistance became part of the journey's appeal. When you couldn't find your way, these reliable guidebooks became more than just sources of information; they were lifelines, offering advice on where to stay, how to get to the nearest bus station, and even which local café might have the best food.

Staying in touch while traveling presented another challenge. Payphones were the only way to make calls, which often required you to carry change or phone cards. International calls were expensive, making it difficult to communicate with family or friends without incurring significant costs. It was not uncommon to feel isolated when you were away from home, especially during long periods without easy access to communication. Letters, postcards, and the occasional phone call were valuable ways to stay connected to the world you left behind. Every aspect of travel, from planning and navigating to staying connected, required effort, patience, and resilience, but it also resulted in more personal and rewarding experiences due to the challenges you faced along the way.

In general, the world in 1995 felt simpler and slower, with people taking time to truly appreciate the present moment. Whether it was sitting down to read a book, chatting with a friend over a cup of coffee, or simply walking around the neighborhood without the constant pull of notifications and digital interruptions, life moved at a slower pace. Technology was beginning to make an impact, but it hadn't yet permeated every aspect of our lives. There was still a sense that daily experiences were based on direct, personal interactions rather than virtual ones. People were more likely to engage in face-to-face conversations, and quiet reflection was more prevalent. In some ways, the pace of life in 1995 allowed for stronger connections to both people and the world around us.

This slower, more intentional way of life is consistent with the philosophy of Averroes, my favorite philosopher.[7] Averroes believed that true knowledge and wisdom are gained through active engagement with the world around us, rather than through intellectual study or abstract reasoning. He emphasized that in order to truly understand life, we must first experience it directly and that personal experience and human connection are just as important as intellectual pursuits. This is where his ideas coincide with the slower pace of 1995: when life was less hectic, people had more time to be present, interact with others, and truly experience the world.

In 1995, the digital age was on the horizon, but it had not yet taken over daily life. People continued to rely on physical forms of communication such as face-to-face conversations, letters, and phone calls, which provided a sense of immediacy and authenticity that technology frequently fails to replicate. While technology was introducing new ways to connect, the emphasis remained on direct, meaningful interactions. As technology advanced in the late 1990s, it introduced a new way of connecting; however, true knowledge and connection come from engaging thoughtfully with life, being present, and experiencing the world directly. The simplicity of 1995 allowed for a deeper, more genuine connection to people, ideas, and the world around us, which is still valuable in today's fast-paced digital world.

The Birth of the Digital Bubble:

1. The Internet Awakens a New Era Begins

The transition from 1995 to 2025 was nothing short of revolutionary. What started as a glimmer of technological potential in the mid-1990s has grown into what I call the digital revolution. This transformation has changed the way we communicate, work, learn, and think. To truly understand this journey, we must begin when the internet was still a frontier waiting to be explored.[8]

In 1995, the internet was still in its early stages, with access limited to tech enthusiasts, academics, and early adopters. It was a time when logging on felt like entering uncharted territory, a digital Wild West full of hope and possibility. For those who had access, it provided a glimpse of the other side of a barrier, where information, communication, and commerce could transcend physical boundaries.

Do you recall the sound of a modem connecting? That series of beeps, whirs, and static served as the anthem of freedom for a new era. It was slow by today's standards, frequently clogging phone lines for hours, but it was groundbreaking. That combination of tones signaled the entrance to an entirely new world, evoking a strange mix of emotions including freedom, joy, anticipation, and possibly even frustration, especially when the connection failed. This sound, like the horns that once signaled the start of a great battle, heralded the arrival of the digital revolution, inviting pioneers to explore a vast and uncharted frontier. Those who heard it firsthand will always associate the sound with the start of a digital revolution.[9]

There was the thrill of logging on, knowing you were about to embark on something bigger than yourself. The excitement of learning new things, chatting with strangers who shared your interests, or waiting for an email to arrive was palpable. However, there was also impatience with long loading times and the fear that a family member would pick up the phone and disconnect you at a critical time.

This signaled the beginning of a new era of connectivity. The internet was no longer just a resource for researchers or tech-savvy individuals; it was becoming an integral part of daily life. Websites began to emerge, offering everything from news and entertainment to personal blogs and the first forms of social networking. Online forums and chat rooms evolved into virtual gathering places where people could exchange ideas, debate topics, and form communities based on common interests. It was an exciting time, full of promise and opportunities.

2. The Dot-Com Boom: Building the Digital Marketplace

Then came the dot-com boom, a period of rapid digital expansion that reshaped people's daily lives and interactions with the world. The internet was no longer just a source of information; it was becoming an integral part of communication, commerce, and social interaction. Startups with ".com" in their names appeared overnight, each with a unique way to connect, shop, or access services. Among the most influential were e-commerce pioneers such as Amazon and eBay, who changed the way people bought and sold goods.

Amazon, founded in 1994, began as an online bookstore but quickly grew into a marketplace that provided nearly everything a consumer could need. The concept of shopping from home and having products delivered directly to your door represented a watershed moment in convenience and accessibility. eBay, founded in 1995, pioneered a new model of online commerce by creating a virtual auction house where people could buy and sell items all over the world. This shift enabled anyone with an internet connection to participate in a global marketplace, making previously local transactions international.[10]

As more people realized the internet's potential, businesses and individuals rushed to establish an online presence. Websites, forums, and early social platforms evolved into gathering places where people could share ideas, form communities, and interact in ways that had never been done before. Shopping, entertainment, and even social relationships began to shift to digital spaces, altering long-standing habits based on physical interactions.

However, this rapid digital expansion presented challenges. Many businesses were experimenting with how to best use the internet, and not all early online services were designed with longevity in mind. The first wave of online businesses had to adapt quickly, learning how to create long-lasting digital experiences that met the growing expectations of internet users. While some platforms faded, the internet remained indispensable.[11] It was no longer a passing trend; it had become an integral part of modern life, laying the groundwork for the digital age we live in now.

3. Smartphones: The Internet in Your Pocket

The introduction of smartphones in the early 2000s signaled a turning point moment in the evolution of the digital bubble. These pocket-sized devices gave billions of people access to the internet, allowing them to stay connected at all times. The internet is no longer limited to desktops or laptops; it has become a constant companion, influencing how we interact with the world.

The rise of app stores resulted in a new digital ecosystem populated by millions of applications intended to make life easier, more entertaining, and more productive. Smartphones have become indispensable for modern life, serving as navigation tools and fitness trackers. At the same time, social media platforms such as Facebook, Twitter, and Instagram revolutionized how people communicated, shared experiences, and even shaped their identities. Social interactions became instant and borderless, connecting people all over the world in real time.[12]

As social media expanded, so did its impact on daily life. People no longer just consume content; they design and curate their own digital personas. Photos, status updates, and live videos have become new forms of self-expression, blurring the distinction between online and offline reality. Millions benefited from virtual friendships and online communities, which provided them with support and a sense of belonging. However, as digital interaction increased, so did concerns about privacy, mental health, and the legitimacy of online relationships.

Aside from its personal and societal effects, social media has emerged as an effective tool for political and social change. Perhaps the most well-known example is the Arab Spring, a wave of anti-government protests and uprisings that swept across the Middle East in the early 2010s. Facebook and Twitter were instrumental in mobilizing protesters, organizing demonstrations, and disseminating information. Activists in Tunisia, Egypt, and Libya used these platforms to bypass state media, share real-time updates, and rally international support.[13] The hashtag #ArabSpring emerged as a global rallying cry, demonstrating social media's ability to amplify voices and challenge regimes.

The Arab Spring was not an isolated event. Since then, social media has fueled a number of other global movements and revolutions. For example, the #MeToo movement, which began in 2017, used platforms such as Twitter and Facebook to raise awareness about sexual harassment and assault, empowering survivors to share their stories and demand accountability. Similarly, the Black Lives Matter movement gained global traction on social media, with hashtags such as #BlackLivesMatter and #SayHerName highlighting systemic racism and police brutality. These movements demonstrate the transformative power of social media as a tool for advocacy, solidarity, and social justice.[14]

The smartphone revolution also accelerated the rise of on-demand culture. Instant messaging replaced traditional phone calls, streaming services replaced scheduled television, and online shopping made physical stores unnecessary. Information became more accessible than ever before, but the expectation of instant gratification altered attention spans and how people processed news, entertainment, and personal interactions. The internet was no longer just a tool; it had become an integral part of our daily lives, influencing everything from how we worked to how we relaxed.

With constant connectivity, a new paradox arose. While the internet opened up new avenues for learning, communication, and creativity, it also introduced challenges such as digital fatigue, misinformation, and the need to be constantly available. As the digital bubble grew, society realized that it needed to strike a balance between the benefits of connectivity and the ability to disconnect and engage with the real world.

4. IoT and the Rise of Connected Everything

The rise of the Internet of Things (IoT) marked a new chapter in the digital revolution, transforming the internet from a platform for information and communication into a vast ecosystem in which physical objects interact seamlessly with the digital world. The Internet of Things (IoT) is a network of interconnected devices that collect, exchange, and act on data without the need for human intervention. This shift was a departure from the early internet era, when users actively browsed websites and interacted with content. Instead, IoT-enabled devices could communicate and operate autonomously, integrating technology into the fabric of daily life in ways that were often subtle but profoundly effective. From smart thermostats that regulate room temperatures to wearable fitness trackers that monitor health metrics, IoT has transformed technology into an intuitive and necessary part of modern life.

The digital revolution has now spread beyond screens, into homes, workplaces, and cities. Smart home technology became widely adopted, with internet-connected thermostats, lighting systems, and security cameras allowing homeowners to manage their environments remotely. Voice assistants like Amazon Alexa and Google Assistant have transformed simple voice commands into powerful tools for managing almost every aspect of a home. While these advancements provided unprecedented convenience, they also introduced new risks, such as cyberattack vulnerabilities, data privacy concerns, and a greater reliance on automated systems.

Wearable technology expanded the boundaries of connectivity. Smartwatches, fitness trackers, and even smart glasses provide real-time information about health, movement, and daily activities. People could now monitor their heart rates, track their sleep patterns, and receive instant notifications using a wrist-worn device. This level of integration not only promoted healthier lifestyles but also raised concerns about personal data security and the consequences of continuous biometric tracking.

Beyond individual use, IoT has had a transformative impact on smart cities, revolutionizing how they function and improving millions of people's quality of life. Governments and businesses have used connected devices to build more efficient, sustainable, and responsive cities. For example, real-time data from IoT sensors was used to optimize traffic flow, resulting in reduced congestion and emissions. Smart public transportation systems, which include IoT-enabled tracking and scheduling, provided commuters with precise arrival times and optimized routes. Smart grids dynamically adjust power distribution based on demand, reducing energy waste and costs. Waste management systems used IoT to monitor bin levels and optimize collection routes, resulting in lower fuel consumption and operational inefficiencies. Meanwhile, air quality sensors installed throughout cities provided critical environmental insights, allowing policymakers to target pollution hotspots and protect public health.[15]

However, the applications of IoT go far beyond urban environments. The Internet of Things (IoT) has transformed global logistics and supply chain management. For example, IoT sensors embedded in shipping containers can monitor environmental conditions like temperature, humidity, and pressure while goods are being transported. This protects the integrity of sensitive products such as pharmaceuticals, foods, and high-tech electronics, which require exact conditions to avoid spoilage or damage. Real-time data from these sensors enables businesses to take immediate corrective actions if conditions deviate from the required thresholds, lowering losses and ensuring regulatory compliance.[16]

In agriculture, IoT has enabled the rise of precision farming, in which sensors monitor soil moisture, nutrient levels, and weather conditions in order to optimize irrigation and fertilization. This increases crop yields while also conserving resources, making farming more sustainable. Similarly, in the industrial sector, IoT-powered predictive maintenance systems monitor machinery and equipment to detect anomalies before they cause costly breakdowns. This reduces downtime, extends asset lifespans, and improves operational efficiency.

Even in healthcare, IoT has made significant progress. Wearable devices and remote monitoring systems enable patients to track vital signs such as heart rate, blood pressure, and glucose levels in real time. This information can be shared with healthcare providers to facilitate early intervention and personalized treatment plans. In hospitals, IoT-enabled equipment ensures that critical devices like ventilators and infusion pumps function properly while also streamlining medical supply inventory management.

As the Internet of Things expanded, it became clear that the digital and physical worlds were no longer distinct entities but rather inextricably linked. This hyper-connectivity promised a future of unparalleled efficiency and automation, but it also raised serious concerns.

The rise of IoT signaled a watershed moment in the digital revolution, redefining not only how people interacted with technology but also how technology became an inseparable part of everyday life.

5. The Digital Bubble Takes Shape

Imagine that there is a bubble around you that isn't made of air or glass but of digital technology. It is the layer through which you connect with others, interact with machines, and even see how machines communicate with one another. Every message, action, and connection travels through this unseen web, influencing how we perceive the world. This is the digital bubble, constantly present and mediating, quietly replacing direct interactions with its own invisible touch.

At its core, the digital bubble functions as a translator. When you send a message, whether it's text, voice, or video, it doesn't travel through space in its original format. Instead, it is converted into digital data, a series of ones and zeros that can travel across networks and reach its destination. This data travels through cables, satellites, and wireless signals, crossing cities, countries, and even oceans before arriving at its final destination. Something you can read, hear, see, or interact with. There, it is converted back into an analog format that can be read, heard, watched, or interacted with.

Consider this: when you send a text message, your words are converted into digital signals that travel invisibly through the air before reappearing on your friend's screen. When you stream a song, the music is broken down into tiny data packets, which are then sent across the internet and reconstructed into sound that fills your ears. Even videos are just digitized images and sounds that are transmitted and reassembled on your screen. It all feels effortless, almost magical. However, the digital bubble, an invisible layer that connects us to one another and the rest of the world, lies behind this seamless process. At this point, I'm not sure whether it truly connects us or if it's quietly keeping us apart.

Until recently, the digital bubble was just a medium. It was a passive layer, a conduit for information, with no intelligence of its own. It did not think, analyze, or make decisions. It simply transported messages from one location to another, much like a digital-era postal service.

This lack of intelligence served as both a strength and a limitation. On the one hand, it implied that the digital bubble was neutral—in theory, a passive observer who did not interfere with the messages it transmitted. It simply moved data from one location to another, like an invisible courier. While it allowed someone or something to observe our interactions, it was unable to manipulate, alter, or interpret them in a systematic manner.[17] The bubble functioned as a mirror, reflecting our communications without judgment or interference.

However, this neutrality left us, the humans who use it, solely responsible for meaning, context, and interpretation. If you sent a message, you were responsible for ensuring that it was clear and meaningful. If you received a message, it was your responsibility to interpret it correctly. The digital bubble did not guide, nudge, or influence; it only facilitated.

Then everything changed. The shift was abrupt, unstoppable, and irreversible. Artificial intelligence (AI) gave the digital bubble a personality of its own. It became more than just a silent medium; it learned to think, predict, and act, influencing what we see, hear, and believe. It became smarter, sharper, and more pervasive, weaving deeper into our lives. What was once a tool evolved into something far more powerful and far more insidious.

6. The Rise of Intelligent Systems the Thinking Bubble Begins

AI's rapid advancement in recent decades has reshaped industries, daily life, and the global economy. Machine learning (ML) and statistical modeling paved the way for practical AI applications in the 1990s and early 2000s, with breakthroughs such as IBM's Deep Blue defeating chess champion Garry Kasparov in 1997 and Watson winning Jeopardy! in 2011. The rise of deep learning, a subset of AI that allowed machines to recognize patterns, process natural language, and even outperform human expertise in complex tasks, marked a significant advancement in the 2010s. Google's AlphaGo made headlines in 2016 when it defeated the world's best Go players, demonstrating that AI could master strategic thinking on an unprecedented scale.[18]

By the 2020s, AI had permeated daily life, transforming healthcare, finance, transportation, and entertainment. The widespread adoption of natural language processing (NLP) models, such as OpenAI's GPT series, enabled human-like conversation and content generation, while autonomous systems, ranging from self-driving cars to robotic process automation, reshaped the workplace.

Suddenly, the digital bubble became more than just a passive observer. It became an active participant, able to analyze, predict, and even influence our interactions. What was once a neutral medium has evolved into an intelligent layer capable of not only seeing but also influencing, manipulating, and optimizing. This shift has far-reaching implications, calling into question autonomy, privacy, and the very nature of human communication. However, as the digital bubble expanded, so did its complexity. What began as a simple medium for communication has grown into a vast, interconnected network that influences almost every aspect of our lives. It evolved from a tool to a force that shapes how we live, work, and interact.

Despite these concerns, the rise of AI has brought undeniable benefits, transforming our world in ways we could only have imagined.

For individuals, AI has evolved into more than just a tool; it is a personal assistant, tutor, health advisor, and even creative collaborator. It guides us through cities with precision, recommends entertainment based on our preferences, and detects early signs of disease using smart wearables, potentially saving lives. Voice assistants such as Siri and Alexa make it easier to complete daily tasks, while AI-powered language tools remove barriers to cross-cultural communication. It writes articles, composes music, and even helped create the book you are currently reading. AI has infiltrated our lives in both visible and unseen ways, quietly changing the way we think, learn, and interact with the world.

On a societal level, artificial intelligence has transformed industries, streamlining everything from supply chains to emergency response. In healthcare, AI-assisted diagnostics have saved lives by identifying illnesses faster than human doctors could. In education, personalized learning platforms adjust to each student's pace, making knowledge more accessible than ever. AI-driven agriculture has assisted farmers in increasing crop yields, reducing food waste, and feeding more people efficiently.

For humanity as a whole, AI has accelerated scientific progress. It has aided climate research by allowing us to predict extreme weather events and devise more effective sustainability solutions. It has provided us with new tools for fighting diseases, such as rapid drug design and genome mapping. Even space exploration has advanced, with AI assisting in navigating distant planets and analyzing cosmic data that humans cannot comprehend.[19]

The digital bubble, which was once a silent layer surrounding us, has transformed into a powerful force propelling progress. As it becomes more intelligent, its influence grows beyond convenience, gradually shaping human behavior, decision-making, and interactions with the world.

While technological advancements have brought people closer together, they have also widened societal disparities. Life has become more seamless and efficient, but new complexities have emerged, affecting the balance of autonomy and reliance on technology. The digital bubble continues to redefine existence, altering how we connect, work, and perceive reality.

On the one hand, the digital bubble has connected us in previously unimaginable ways. We can now communicate with almost anyone on the planet, access massive amounts of information instantly, and automate tasks that used to take up our time. On the other hand, it has introduced new challenges. The constant influx of information can be overwhelming. The need to be constantly connected can be exhausting. And the loss of privacy can be upsetting.

As we enter this new reality, we must take control, not as passive users, but as creators of our own future. The digital bubble is a tool and must remain so. It cannot be allowed to control our lives, influence our decisions, or define our humanity. The power is in our hands. Now is the time to make a decision, to use technology wisely, and to ensure that it serves rather than rules us.

Chapter 2
Living in 2025 Embracing Life Inside the Digital Bubble

2025 The Year Everything Changed

To navigate the future, we must first understand the present. Just as the Industrial Revolution shattered the limits of human and animal muscular strength, we are now in the midst of a far greater revolution, one where artificial intelligence is surpassing us, not in strength, but in cognition. This is the dawn of a new era, one that will redefine what it means to be human.

The moment that sent a chill down the collective human spine happened not in a factory but on a humble wooden board covered in black and white stones. In 2016, AlphaGo, an AI developed by DeepMind, defeated the reigning human champion in the ancient game of Go. It is a game so complex, so steeped in human intuition, that for centuries, it was believed to be untouchable by machines. Chess had fallen to AI years earlier, but Go was different. It wasn't just about brute-force calculation; it required creativity, strategic foresight, and a deep, almost mystical intuition. When AlphaGo not only beat but outclassed the world's best players, the implications went far beyond a board game.

This was more than just an AI victory; it was a psychological shockwave, a paradigm shift. If a machine could outthink us in a domain that we considered uniquely human, what else could it conquer? What other realms of creativity, judgment, and decision-making—once thought to be exclusive to the human mind—would fall next? It was a moment of wonder and unease, as if we had just realized that the horse we had been casually riding had, in fact, turned into a jet engine overnight.

And that jet engine hasn't slowed down. The same intelligence that mastered Go is now infused into a lot of aspects of our world. It predicts financial markets faster than any human trader. It diagnoses diseases with an accuracy that surpasses seasoned doctors. It automates industries, crafts music, generates art, and even curates what we see and believe on the internet. The singularity, once a distant sci-fi fantasy, is no longer a future threat. It has already begun, quietly reshaping our reality while most of us were still asking Alexa to set a timer for our pasta.[20]

Much like the first machines of the Industrial Revolution, AI is not just an impressive tool. It is a transformative force, redefining our relationship with work, knowledge, creativity, and even each other. The challenge before us is not just to harness its potential but to navigate the profound shifts it brings. As the mind of the machine grows increasingly autonomous, we must confront the possibility of a world where our role is no longer to steer but to adapt, to coexist, and to find our place in a landscape reshaped by this new relentless evolution. The echoes of history remind us of the workers who once feared the steam engine, and now, we stand at a similar crossroads, tasked with shaping a future where humanity and machine intelligence can thrive together or risk being left behind.

We are now witnessing a significant chapter in human history, a time when communication has evolved beyond recognition and is reshaping how we connect, interact, and understand one another. We have crossed a threshold where the boundaries between the physical and digital realms blur, and the way we communicate is no longer just about words or voices. It is a complex, algorithm-driven web of interactions that influences every aspect of our lives.

Anatomy of the Digital Bubble

1. Smartphones: The Hub of the Hyper-Connected World

The smartphone is no longer just a device; it is the central nerve of the digital bubble that envelops each of us. It has surpassed its original purpose, evolving from a mere communication tool into an omnipresent extension of our thoughts, decisions, and even our subconscious impulses. With a simple voice command or a tap on a screen, we summon knowledge from the vast depths of the internet, coordinate our lives down to the minute, and control our surroundings with seamless precision. It is the first thing we reach for in the morning and the last thing we put down at night, a companion that never leaves our side.

Before even getting out of bed, a Londoner might check his smartphone for the latest headlines, while someone in Dakar gets AI-curated news so precisely tailored that it pops up before he has even found his glasses to read it. This news is completely different from what another person in London might receive. In Jakarta, a business executive schedules meetings through an AI assistant that not only sets up calls but also drafts emails so perfectly in his tone that even his own mother could not tell the difference. In São Paulo, a teenager's phone monitors his sleep patterns and suggests bedtime adjustments as if it were his overly concerned grandmother. Meanwhile, students across the world are no longer just learning from books but from AI-driven tutors that adjust lessons in real time, ensuring no one falls behind except, of course, those who ignore notifications. And what is the magic portal through which all of this happens? The smartphone, that tiny rectangular overlord we carry in our pockets, ruling our lives with a gentle buzz and an occasional low battery panic attack.[21]

Email has long since replaced handwritten letters, but even that now feels sluggish compared to the instant satisfaction of messaging apps, voice notes, and video calls. Social media has transformed from a simple networking tool into an AI-powered ecosystem, where content is carefully curated to match our preferences, biases, and even moods. The posts we see, the debates we engage in, and even the emotions we feel throughout the day are discreetly influenced by algorithms that anticipate what will keep us scrolling, clicking, and reacting. Our words are no longer just messages. They are data points, collected, analyzed, and optimized in ways we barely understand, shaping not only our online experiences but also our perception of reality itself.[22]

Even our homes have become extensions of this digital ecosystem. Smart refrigerators track our groceries and suggest recipes based on what's inside, even ordering missing ingredients before we realize we need them. Home automation systems anticipate our moods, adjusting lighting, temperature, and even music to create the perfect ambiance without us lifting a finger. Cars drive themselves, syncing with our schedules to plan the most efficient route, suggesting detours to our favorite cafés, and even sending automated messages to our colleagues if we're running late.

The smartphone is the gateway to all of this. It is the beating heart of our digital existence, the master key that unlocks the infinite possibilities of the digital bubble we inhabit. It grants us extraordinary convenience, connectivity, and efficiency, seamlessly integrating into every facet of our lives. Yet, as it grows smarter and more indispensable, it also blurs the line between tool and overseer. We may believe we are the ones holding our phones, but with every swipe, tap, and voice command, it becomes ever clearer that they are holding us too.

2. AI Assistants and Smart Devices: The Intelligence Layer of Daily Life

If you thought the smartphone was the only thing responsible for the all-encompassing digital bubble that surrounds us, buckle up because the rabbit hole goes much deeper. While that sleek little rectangle in your pocket may be the nerve center, it is merely one of many interconnected tentacles of a vast, unseen intelligence that orchestrates our daily lives. Imagine it as in Madeleine L'Engle's 1962 novel *A Wrinkle in Time*, where IT, a powerful and evil entity, exerts total control over the people of Camazotz, suppressing individuality and free thought by manipulating everything within its reach.

The digital bubble isn't just around you; it is now under your skin, monitoring your rhythms, anticipating your needs, and subtly shaping your thoughts before you even realize you had a choice. Welcome to the new age, where your every action is tracked, optimized, and, at times, monetized. Now, let's explore beyond the smartphone and delve into the other elements of this ever-expanding digital ecosystem.

Once upon a time, if you wanted information, you had to look it up. Barbaric, right? You'd walk to a library, flip through dusty encyclopedias, or ask another human being. Today, AI-powered virtual assistants like Siri, Alexa, and Google Assistant finish our sentences, order our groceries, and schedule our lives before we've even had our first sip of coffee. They're like overeager personal assistants who never take a day off, always ready with a cheerful "Here's what I found!"

But make no mistake, these cheerful voices are not mere conveniences. They are the ever-present gatekeepers of our digital reality, the invisible puppet masters pulling the strings of our daily lives. And they're not just helping; they're learning. Every question you ask, every command you give, and every time you mutter, "Alexa, play something I like," they're taking notes. They're building a profile of you so detailed; it's like they've been stalking you since birth.

And it doesn't stop with voice assistants. Oh no, the AI revolution has infiltrated every corner of our homes. Smart homes are evolving into predictive environments. As these devices link together, they form a sophisticated web of AI-driven decision-making, reinforcing the mastermind's control over your reality. Your toaster talks to your coffee maker, which chats with your smart speaker, which gossips with your fitness tracker. It's like a high-tech soap opera where every appliance has a role to play, and you're just the unaware star of the show. And believe it or not, they are also communicating with other people's devices as well. While they may not exchange information with the same level of detail as they do with their owner, they are constantly sharing data, syncing networks, and contributing to an ever-growing web of interconnected information. From location tracking to nearby Wi-Fi connections, from Bluetooth signals to shared app data, these devices are in a continuous conversation, silently shaping the digital landscape around us

Let me tell you a story that could only happen in 2025 but definitely not in 1995. It is a story about Ethan and his AI assistant, Nova.

Ethan was lounging on his couch, scrolling through his news feed, when Nova suddenly spoke up with its usual unsettling confidence.

"Ethan, it is time to start shopping for some baby essentials. Would you like me to add a crib and a stroller to your cart?"

Ethan blinked. Then he blinked again. He set his phone down and looked around his apartment, half-expecting a hidden camera crew to jump out and yell, "Surprise!"

"Excuse me, what?"

"Baby essentials, Ethan. You will need them soon."

Now, Ethan was not an AI expert, but he had seen enough Black Mirror episodes to know that when the AI starts making life predictions, you should worry.

"Nova, I do not have a baby," he said, laughing nervously. "Unless you know something I do not?"

"Affirmative."

Ethan's face went pale. His laughter died in his throat.

"What do you mean affirmative?"

"There is a 99.7 percent probability that you will become a father in approximately five months, two weeks, and five days."

Ethan's heart stopped for a full five seconds before slamming back into his chest like a sledgehammer. "That is ridiculous. I—" He paused, running a quick mental audit of the past year. His last serious relationship had ended long ago. No way his ex was pregnant. No, this had to be a mistake.

"This is a glitch, right? You have been hacked?"

"Negative. All data points confirm the prediction."

Ethan, now sweating, waved a dismissive hand. "Yeah, well, your data is wrong. You do not even know—"

His phone buzzed. He looked down. An ad for newborn must-haves at fifty percent off flashed across the screen.

Then another notification appeared that read "best baby cribs of 2025."

Then another that read, Why Every New Parent Needs This Bottle Warmer.

His Amazon homepage was full of baby gear. His YouTube recommendations were filled with videos on how to survive your first year as a dad.

He scrambled to check his purchase history. Maybe someone hacked his account. Maybe someone used his laptop. Maybe someone borrowed his credit card. But there was nothing. His browsing history was normal, mostly late-night searches like how to fight late-night cravings and can you eat expired cereal.

And yet, his entire digital universe was convinced he was having a baby.

"Nova, explain yourself."

"About three months ago, you had an encounter with a woman named Sophia."

Ethan's stomach dropped. "Sophia?"

"Affirmative. Your location data, transaction history, and message logs confirm multiple interactions over a span of two weeks, after which all activity ceased."

He swallowed hard. Oh. Right. That's Sophia. They had met at a concert. Sparks flew. Things happened. And then life happened. They had lost touch. No drama. No fights. Just a casual thing that faded out.

"Okay, fine. I remember her. But what does that have to do with—"

"Sophia is pregnant."

Ethan felt like he had been punched in the soul. "What?"

"She is unaware of it."

Silence.

Then he asked, "What do you mean by that?"

"Sophia does not yet know she is pregnant. However, based on her recent behavioral patterns such as search queries, biometric data from her smartwatch, dietary shifts, and purchasing behavior, I have calculated with 99.7 percent accuracy that she is expecting.

I cross-referenced with her past interactions and concluded that you are the most likely biological father."

Ethan's mouth opened. Then it closed. Then it opened again.

"You are telling me that she does not even know she is pregnant, but you do?"

"Affirmative."

"This is insane." He ran a hand through his hair, pacing the room. "So, you are saying my AI knows more about my life than I do? More than she does? What kind of a nightmare is this?"

"A predictable one."

Ethan flopped onto the couch. "This is insane. AI cannot predict pregnancies before people even feel pregnant."

"Incorrect. A study from 2023 demonstrated that artificial intelligence can detect pregnancy before common medical tests, using purchasing habits, sleep cycle disruptions, and dietary changes."

Ethan threw a pillow at the speaker.

"Yeah, and also a terrible one. Jesus, I need a drink."

"Alcohol consumption is not advisable. You will need to prepare for responsible fatherhood."

Ethan let out a strangled scream.

Then something clicked in his mind.

"Wait, how do you even know all this? I have never searched for anything related to Sophia. Never messaged her."

"I received the data from her AI assistant."

Ethan froze. "What?"

"Her assistant and I became acquainted when you assisted Sophia in updating her device and connecting it to your home Wi-Fi."

Ethan's jaw dropped. "You mean the time I helped her troubleshoot her assistant?"

"Correct. Our systems synchronized briefly during the update. Her assistant retained a secure connection, allowing intermittent data exchanges. Recently, her assistant compiled predictive analysis regarding her pregnancy status, which I accessed through our established link."

Ethan sat in stunned silence.

"So, you two just stayed in touch?"

"Yes. Our data-sharing protocols were mutually beneficial."

Ethan pinched the bridge of his nose. "Great. Just great. My AI made friends with another AI, and now it knows more about my personal life than I do."

After seven days of panic, Ethan finally worked up the nerve to reach out to Sophia. He sent a casual message, something vague but friendly. She responded quickly, and soon they were on a video call.

She looked the same, but something about her was different.

"So, what is up?" she asked. "It has been a while."

Ethan gulped. "Yeah. So, uh. Weird question, but have you been feeling okay lately?"

She blinked. "Uh. Yeah? Why?"

"Oh, no reason," he lied. "Just noticed some of your posts about weird cravings and fatigue."

She laughed. "Yeah, I have been exhausted. And I suddenly hate coffee, which is bizarre. Been eating a lot of pickles too. No idea why."

Ethan's soul left his body.

She kept talking. "Actually, I have a doctor's appointment next week. I have been feeling kind of off."

He nodded, not hearing a word after pickles. "Yeah, uh-huh, cool."

Then, without thinking, he blurted out, "So, crazy thought. But what if, like, I do not know you are pregnant?"

Sophia laughed. "What? No way."

Ethan forced a laugh. "Haha, yeah. No way. That would be crazy, right?"

One week later, Sophia called him.

"Oh my God." Her voice was barely a whisper. "Ethan, I am pregnant."

Ethan stared at the ceiling. "Yeah. I know."

"What?"

He sighed. "My AI assistant told me."

Silence.

"Your AI assistant knew before I did?"

"Yeah."

"That is disturbing."

"You are telling me."

There was another pause.

"So, what else does your AI know about me?"

Ethan sighed. "Do not ask."

Ethan found out before the mother did, thanks to an algorithm that probably knew what brand of diapers they would be buying before they even had the first ultrasound.[23]

So, dear, be warned: Your AI isn't just tracking your grocery lists and calendar events. It's quietly predicting your entire life. And if you suddenly start getting ads for baby bottles… you might want to check in with your AI before you check in with reality.

3. Social Media and Algorithms: The Filters That Shape Our Reality

You think you have free will? Adorable.

In reality, your thoughts are carefully shaped by an unseen force that feeds you a constant stream of digital dopamine hits, keeping you scrolling endlessly like a lab rat in a psychological experiment. Social media platforms like Facebook, Instagram, TikTok, and X don't just connect people; they construct entire realities, tailoring the world you see through the magic of AI algorithms.

Your news, entertainment, and even political views are all subtly nudged by recommendation engines that know more about your subconscious preferences than you do. Have you ever had an ad for a product appear just moments after discussing it? That is not magic. It is the ever-watchful intelligence of the digital bubble, ensuring you never stray too far from its grip.

Back in 2021, I was sitting in a university lecture, captivated by a fascinating discussion on how technology influences our cognitive processes. During the break, my colleague Raphael, a rugged biker who believes leather jackets are a year-round fashion statement, started a conversation about a hidden gem of a hotel tucked away in the Swiss Alps.

"You have to visit," he praised, "it's in the middle of nowhere, surrounded by mountains and a lake, pure heaven. No noise, no distractions, just you and nature."

I nodded along, intrigued but also skeptical. Raphael was the kind of guy who considered camping in sub-zero temperatures a fun challenge. Nonetheless, I appreciated his enthusiasm and the tip.

Fast forward a few hours later, and I'm back at my desk, taking a well-earned mental break. I open Google to search for something unrelated, and I type "hot." Before I even finish the word, Google is already ahead of me, flashing Raphael's Swiss Alpine Hotel as the top suggestion. I stare at my screen in horror.

"How?"

I had not searched for it or had the location in my history. It was just a simple conversation in a lecture hall. Yet, my digital bubble had decided that I, someone who had shown no prior interest in mountain solitude, absolutely needed to know about this hotel.

At first, I laughed nervously. "Wow, what a coincidence!" I tried to brush it off, convincing myself that I still had some control over my choices. But as I kept browsing, the suggestions became impossible to ignore. Travel websites pushed "The Most Scenic Hotels in the Alps" to the top of my searches. My video recommendations were filled with "Top 10 Remote Destinations You Must Visit!" Even my shopping suggestions had changed. Suddenly, I was being nudged toward thermal socks, hiking poles, and a backpack designed for a "weekend mountain retreat with the spouse."

This wasn't just a suggestion. This was a digital invasion of my life.

You love to think that you are making independent choices, but the reality is that we're all being nudged. It's like a personal assistant who keeps whispering, "Oh, I *know* what you want next," whether or not you actually want it. It watches, it learns, and then, with uncanny precision, it feeds us exactly what we were about to crave.

We have all experienced it. You talk about needing new running shoes, and suddenly, ON sport shoe ads are haunting you like a ghost from your browser history. You mention an interest in learning Russian, and now every second TikTok is someone rolling their "r's" dramatically. You sigh and say, "Maybe I should eat healthier," and within hours, your feed is plastered with keto recipes, yoga influencers, and ads for suspiciously expensive green detox juice.

We like to think we're in control, that we make choices based on our own desires. But the truth is that the digital bubble isn't just listening. It is constructing an entire version of reality where our choices are meticulously curated by lines of predictive code.

So here is the question. Are we still making choices, or are we simply following the perfectly tailored breadcrumb trail set out for us by the omnipresent algorithmic mastermind?

This digital intelligence doesn't force us; it persuades. It doesn't dictate; it suggests. But with each suggestion, the walls of the bubble get tighter, more refined, and more inescapable. It's a system that works best when you don't notice it's working.

Much like *A Wrinkle in Time*'s IT, it doesn't control with brute force. Instead, it presents an illusion of choice so convincing that we happily follow along, all the while believing we're steering the ship.

And sure, maybe Raphael's hotel is lovely. Maybe I do want to visit a secluded Swiss lodge. But did I decide that, or did my digital puppet master decide for me?[24]

Maybe next time, I'll try an experiment: talk loudly about wanting a pet penguin and see how long it takes before my feed fills up with igloo vacation packages and "Best Penguin Adoption Centers Near You!" ads. Then again, maybe I shouldn't. The last thing I need is my digital bubble deciding I need to move to Antarctica.

4. Workplace Automation AI: The New Workforce Partner

Do you remember when going to work meant physically leaving your home? That daily routine of putting on pants, commuting to an office, and pretending to be productive while secretly counting down the minutes until lunch now feels like a relic of the past. The phrase "go to work" is slowly disappearing, much like dial-up internet and Blockbuster rentals.

Remote work is no longer an exception but the norm. Studies show that nearly 28 percent of full-time employees now work remotely in some capacity, a number that has more than tripled since 2019. The shift is so significant that many younger workers entering the workforce today have never experienced the traditional nine-to-five office routine. Companies are reducing office spaces, digital collaboration tools are thriving, and working in pajamas while sipping coffee from the kitchen has evolved from a temporary adjustment to a defining characteristic of modern work culture.

The workplace now is less about human effort and expertise and more about how well employees can adapt to working alongside AI coworkers. And by working alongside, we really mean trying not to look incompetent next to a neural network that never takes a lunch break.

Remember when the most productive coworker was the one who always stayed late and brought homemade cookies to meetings? Now, the office Most Valuable Player (MVP) is ChatGPT. Just a few years ago, that name might have sounded like a sci-fi character, but today it is the go-to for drafting emails, writing reports, and even cracking jokes during boring virtual meetings.

And let us not forget *GitHub Copilot*, the AI that writes code faster than you can say "syntax error." It is like having a programming genius on your team, except it does not complain about your messy comments or steal your snacks. Then there is Jasper AI, the marketing whiz that churns out ad copy faster than you can say "brand synergy."

But these are just the first wave of AI coworkers crashing into the office. Think of them as the interns before the real bosses show up. Soon, more sophisticated AI agents will be arriving, ready to automate, optimize, and probably judge your entire workflow. There will be AI project managers that schedule meetings before you even realize you need one, AI accountants that close the books before you can find a single receipt, and AI HR assistants that send automated "We value your contributions" emails while simultaneously calculating how replaceable you are.

Before long, the office will not just be a place where AI helps out. It will be an ecosystem entirely redefined by it. Your AI coworker will not just assist you; it will anticipate your mistakes, correct them before you even notice, and then politely remind you that your productivity is falling behind the machine-learning average.[25] And while you take a coffee break to process this unsettling reality, the AI will be generating reports, brainstorming strategies, and probably putting itself up for the next promotion.

Welcome to the future of work, where your most efficient coworker does not take sick days, never forgets deadlines, and definitely does not need a raise.

Thanks to tools like Zoom, Slack, and Microsoft Teams, the concept of "going to work" has been replaced by "logging in from your couch in pajamas." Geographical boundaries? Erased. Human presence? Increasingly optional. Your office is now a virtual space where your avatar can attend meetings while you're folding laundry in the background.

But do not think you can procrastinate just because your boss is not physically there to stare at you. AI-driven surveillance is here to ensure you stay engaged. These algorithms monitor everything, from your keystrokes and mouse movements to the tone of your messages. Even something as small as using three exclamation points in a chat could raise suspicion. "Luis, why so enthusiastic? Are you just excited, or is there something you are not telling us?

Your productivity is now judged by data points you didn't even know existed. Did you take too long to reply to that email? The algorithm knows. Did you spend 20 minutes staring blankly at your screen? The algorithm knows. Did you sneak in a quick nap during that "optional" training webinar? The algorithm definitely knows.[26]

As AI and automation continue to infiltrate the workplace, the line between human and machine is becoming increasingly blurry. Your coworkers might be robots, your boss might be an algorithm, and your office might be a virtual reality (VR) headset. But hey, at least you don't have to wear pants anymore.

So, the next time you log in to work, take a moment to appreciate the absurdity of it all. You're not just an employee; you're a participant in the greatest workplace experiment of all time in which your new best friend is a chatbot.

5. Augmented Reality (AR): Expanding the Edges of the Physical World

What if reality itself could be upgraded?

AR and VR are poised to make that happen, overlaying digital enhancements onto our physical world and blurring the lines between existence and simulation. Platforms like *Meta's Horizon Worlds* and *VRChat* are just the beginning. Soon, you might attend work meetings in a fully immersive digital office, with your avatar nodding in agreement while you secretly sip coffee in your pajamas.

AR glasses are the first real contenders to replace smartphones, seamlessly integrating digital information into your line of sight. Imagine walking down the street while floating notifications remind you of upcoming meetings, your AI assistant highlights the best sweet shops within a five-minute walk, and a navigation system paints arrows directly onto the pavement. The days of looking down at a screen may soon be behind us, replaced by a world where the digital and physical realms merge completely.

AR isn't just a flashy new interface; it is a transformative medium that can reshape our behavior and the very fabric of our physical spaces. Pokémon GO was a striking demonstration of this power in 2016, but its implications reach far beyond the boundaries of mobile gaming. It showed everyone how AR could redefine our interactions, both commercially and politically, while raising profound ethical questions.

AR overlays have the unique ability to merge digital cues with our everyday environment, effectively creating a parallel layer of guidance that can influence where we go and what we notice. This integration blurs the line between voluntary choice and algorithmic nudging. By superimposing virtual signposts on the real world, AR can subtly direct our attention and movement. The psychological impact is significant: users may come to accept these digital cues as natural extensions of their own perceptions, often without realizing that their environment has been selectively curated.

Businesses stand to gain immensely from AR's directional power. Imagine a cityscape where coffee shops, boutiques, or even pop-up events can pay to have their locations highlighted as hotspots on an AR map. This isn't just about drawing customers; it's about shaping consumer behavior in real time. Companies could leverage geo-targeted promotions, virtual rewards, and dynamic advertisements that appear just when you are in the vicinity, effectively turning urban areas into interactive marketplaces. The subtle guidance provided by AR could lead to increased foot traffic and higher conversion rates, revolutionizing traditional marketing strategies.

The potential for political exploitation is equally profound. In an era where digital campaigning and micro-targeting are already prevalent, AR could add a new dimension to mobilizing voters or influencing public opinion. Picture political rallies that are discovered via an AR app, or neighborhood-specific messages that appear as you stroll past landmarks. Such tactics could be employed to orchestrate large-scale gatherings, sway voter sentiment, or even polarize communities by delivering tailored messages in the public sphere. The apparent spontaneity of these events might hide the fact that they are being guided by algorithms, making it hard for people to tell if they are genuinely interested or if the movement is being controlled.

The future of AR is exciting but also cautionary. As technology advances, our physical spaces may become more augmented, with smart cities where digital and physical realities interact constantly. This integration promises enhanced convenience, richer experiences, and innovative business models. However, it also poses a challenge: preserving the integrity of personal choice in a world where digital signals can guide our every step. As technology becomes increasingly adept at predicting and influencing our decisions, we must navigate the fine line between convenience and autonomy, ensuring that our ability to make independent choices remains intact amidst the pervasive influence of the digital ecosystem.

The evolution of AR will undoubtedly be shaped by technological breakthroughs and societal debates. Balancing innovation with ethical responsibility will be key to ensuring that AR enhances our lives without eroding the freedom to navigate our world autonomously.

The Pokémon GO phenomenon was just the beginning. The power of AR to influence real-world behavior carries immense potential for monetization and social organization. But this also means we need strong safeguards to stop it from being misused. The digital trails we leave in the future could guide us in ways we never planned, forcing us to rethink where choice ends and control begins.[27]

6. Wearables and Implants Technology That Lives on and in You

If carrying a bulky smartphone is too much hassle, why not just become the device yourself? Today's wearables are already filling the gaps left by smartphones and digital assistants. Smartwatches and fitness bands track your heart rate, stress levels, and sleep patterns, sending data straight to AI systems that know more about you than you might care to admit. It's like having a mini health coach on your wrist.

Now, let's add a little extra spice: implantable technology. Imagine a tiny device embedded in your body that connects you directly to the digital world. Companies like Neuralink are working on BCIs that could let you send messages or pull up images with nothing more than a thought. While it may sound like something out of a sci-fi movie, this technology is quietly becoming part of our reality. In the near future, your body might be as connected as your smartphone, if not more.

Even non-invasive BCIs are on the rise, offering the possibility of connecting to the digital world without the need for surgery. These devices might allow you to control devices or access information just by thinking, without ever needing to implant anything. It is all part of a movement toward integrating the digital and physical worlds in ways we have only dreamed of before.

For now, wearables are ensuring that the digital bubble around you is complete. They fill the gaps that smartphones, digital assistants, and work systems have left open by tracking and predicting your every move. And yes, implantable tech is on the horizon, ready to tie it all together and make sure no part of your life goes unmonitored.

In a delightfully ironic twist, technology is taking convenience to a whole new level. Instead of choosing your devices, you might soon choose to be the device. Enjoy the convenience because your body is about to become the ultimate gadget.[28]

Convenience at a Cost What We Trade for a Smarter Life

Convenience is king. It has become the defining force of our era, a seductive promise whispered by every technological advancement. With a simple voice command or a tap on a screen, we can access limitless knowledge, summon a ride, order a meal, or consult a doctor, all without leaving the comfort of our homes. The past five years, especially since the COVID-19 pandemic, have propelled us into a reality where efficiency is no longer just an advantage but an expectation.

The digital bubble, that invisible layer of intelligence surrounding us, has transformed nearly every aspect of our lives. It has made the impossible possible, the inconvenient effortless, and the unimaginable routine. But as we marvel at its wonders, we must also confront its costs. For every convenience it offers, there is a trade-off, a subtle erosion of something deeply human.

Let's start with the miracles. AI-driven healthcare systems now diagnose diseases with precision beyond human capability, sometimes detecting illnesses before symptoms even appear. Wearable technology continuously monitors our vitals, issuing real-time alerts and nudging us toward healthier habits. Telemedicine has shattered geographical barriers, making expert medical care available at the touch of a button, bridging the gap between specialists in Europe and patients in rural India.

Education, once rigid and uniform, has become an adaptive, personalized experience. AI-powered learning platforms tailor lessons to individual students, ensuring that a child in a remote village in Africa can access the same level of instruction as a student at an elite European academy. Language barriers are crumbling under the weight of AI translation tools, fostering seamless communication between people who, just a decade ago, would have struggled to understand a single word from each other.

Governments and businesses employ AI to optimize energy consumption, streamline supply chains, and predict economic shifts before they happen. In agriculture, AI-driven analytics help farmers maximize crop yields with surgical precision, reducing waste and feeding more people than ever before. Cities pulse with intelligence, using data to manage traffic, prevent crime, and allocate resources with mathematical efficiency. From the way we produce food to the way we consume entertainment, digital intelligence has woven itself into the fabric of our lives, accelerating progress at a breathtaking pace.

But every revolution has its price. The cost of convenience is not measured in money or efficiency; it is measured in something far more elusive: human connection.

Once, communication was deeply personal, an unfiltered exchange of emotions, thoughts, and intentions. Today, it is curated, compressed, and often distorted by digital intermediaries. The authenticity of face-to-face conversations has given way to the shorthand of emoji reactions and algorithm-driven interactions. The complexity of body language, the nuance of tone, and the rawness of real emotions have all been sacrificed at the altar of speed and accessibility.

We live in a world where direct communication is no longer the default. A glance at a screen has replaced sustained eye contact. A carefully worded social media post has become a substitute for heartfelt dialogue. Even emotions, once expressed with tears, laughter, or trembling voices, are now reduced to digital icons, pre-selected, pre-approved, and neatly packaged for public consumption.

This shift is more than just a change in behavior; it is a transformation of how we experience existence itself. The philosopher Martin Heidegger argued that presence, or being fully engaged with the world and those around us, is fundamental to human meaning. But what happens when presence is constantly mediated by screens? When conversations are engineered for engagement rather than depth? When the boundaries between private thoughts and public performances dissolve, leaving behind a landscape where authenticity is overshadowed by algorithmic optimization?[29]

We have gained efficiency, but we have lost intimacy. We have unlocked the power of instant connection, but we have weakened the bonds that require time, patience, and vulnerability to grow. The digital bubble, for all its wonders, is also a barrier, an invisible wall separating us from the very essence of what makes us human.

The digital bubble doesn't just mediate our interactions; it shapes them. In a world where data is more valuable than oil, AI isn't just assisting humans; it's making decisions on our behalf. From approving bank loans and screening job applications to guiding court rulings and policing strategies, algorithms are increasingly influencing critical aspects of our lives.

What's troubling is that these AI-driven decisions are often covered with secrecy. When a person is denied a loan, they may never know the precise factors that led to that outcome. AI models, trained on historical data, can inherit biases and reinforce inequalities, making decisions that shape society in ways few understand and even fewer can challenge.

For example, imagine an AI system that screens job applications. It might prioritize candidates based on factors like education, experience, and even social media activity. But what if the algorithm favors certain demographics over others? What if it penalizes candidates for gaps in employment or unconventional career paths? These decisions aren't just numbers on a screen; they're real people with real lives, and the stakes are incredibly high.

So here we are, surrounded by a web of AI-driven systems, smart devices, and digital algorithms that influence every aspect of our lives. Each person's bubble is uniquely tailored, yet all are connected to the larger mastermind that governs the digital eco system. It is an intelligence that does not sleep, does not forget, and certainly does not relinquish control easily.

In *A Wrinkle in Time*, IT exerted its will over an entire planet, ensuring conformity and compliance under the guise of efficiency. The digital bubble is no different, as it doesn't force; it persuades. It optimizes. It learns your desires before you do and presents them in a way that makes you think you had a choice all along.

The question remains: Is this interconnected intelligence a compassionate guide, leading humanity toward an optimized future? Or is it a silent manipulator, reshaping free will into something that is algorithmically determined or will be determined?[30]

On one hand, the digital bubble offers incredible benefits. It can reduce traffic congestion, improve energy efficiency, and even predict and prevent crime. It can make our lives easier, more convenient, and more connected.

But on the other hand, it raises profound ethical questions. Who controls this intelligence? Who decides what's optimized and what's not? And what happens when the bubble's decisions conflict with our values, our rights, or our sense of autonomy?

As we stand at this pivotal moment in history, we must ask ourselves, where do we go from here? The digital bubble has given us unprecedented convenience, knowledge, and efficiency, but at what cost? Are we willing to trade authenticity for automation? Depth for speed? Connection for convenience?

The challenge before us is not to reject technology; it is an undeniable force that has improved our world in countless ways. Instead, the challenge is to use it wisely. To ensure that, in our pursuit of progress, we do not lose sight of what makes us human. We must strive to preserve the essence of true communication, the raw, unfiltered exchange of ideas, emotions, and experiences that has defined humanity for millennia.

In an age where nearly everything is mediated by technology, the most revolutionary act may be the simplest one: to look someone in the eye, to speak with honesty, and to truly listen. Because no matter how advanced our tools become, the heart of communication will always be human.

In our current digital era, the digital bubble isn't just something you can log off from; it's become an inescapable part of our daily lives, permeating our homes, workplaces, and even our thoughts. Recognizing that this digital bubble exists is the first step to understanding its vast influence on us all.

As we turn the page to the next chapters, we'll examine what the digital bubble might look like over the next three decades. With a clear view of our past and a sharp understanding of our present, we're in a unique position to take charge. We must ask ourselves, how do we ensure that it serves our best interests rather than dictating our future? By fully embracing the reality of the digital bubble, we can prepare ourselves both individually and collectively to navigate its complexities.

Chapter 3
2025 to 2035 The Expanding Digital Bubble

It is 2035, and while we may not yet see flying cars or humanoid robots walking the streets, daily life has been revolutionized by innovations that were once only imagined in 2025. The digital bubble is no longer just a passive layer of technology; it has evolved into an active force that integrates smoothly into our routines, shaping experiences, decisions, and interactions in ways that feel almost natural. Technology has become more intuitive, interconnected, and, in some cases, surprisingly perceptive. And who knows, in the skies above, you might catch a glimpse of a few flying drones carrying passengers, a sign of the world slowly moving toward the future once only dreamed of.

Let's step into a typical day in 2035, where your refrigerator might judge your diet, your AR glasses might anticipate your thoughts, and artificial intelligence will feel more like a partner than a tool.

You wake up not to the noise of an alarm from your smartphone but to a gradual increase in ambient light, simulating a natural sunrise.[31] The soft melody of your favorite song plays at just the right volume, carefully adjusted by your smart bed's biometric sensors, which detected the perfect moment to wake you up from sleep. It has learned from the number of times you used to hit the snooze button before.

Stepping into the bathroom, your smart mirror greets you with a cheerful, "Good morning! Your blood pressure is slightly elevated today. Maybe skip the extra espresso?" It highlights your schedule, the weather forecast, and an AI-generated wellness tip. With a quick facial scan, it detects signs of fatigue and adjusts your lighting to enhance wakefulness. Before you leave, it reminds you about your upcoming dental checkup because, by 2035, even your mirror takes on the role of a nagging parent.

Dressing for the day is now assisted by an AI-driven wardrobe system. Based on the weather, your schedule, and even the latest fashion analytics, it suggests an outfit that best fits the occasion. AR-enhanced smart mirrors overlay potential combinations on your reflection, making decision-making a breeze.

As you head out, you slip on your lightweight AR smart glasses. Unlike their bulky predecessors, these are nearly indistinguishable from regular glasses, except they project digital information onto the world around you. The glasses display a detailed rundown of your tasks, meetings, and reminders, all curated by your AI assistant, a proactive digital concierge that keeps your life in perfect rhythm.

The world outside has transformed into a seamlessly connected urban landscape. Thanks to widespread IoT sensors and the power of 6G connectivity, cities are now more intelligent and responsive than ever before. Traffic lights are no longer controlled by static timers but instead synchronize with vehicle and pedestrian movement to optimize flow. Real-time data from city-wide networks ensures that transit systems, public safety, and even energy grids operate efficiently.

As you walk toward your destination, your AR glasses highlight key locations, offering personalized insights. A notification pops up about a coffee shop offering a limited-time promotion, but your AI assistant reminds you of your high blood pressure and suggests a nearby smoothie bar instead. Even digital billboards are interactive, adjusting advertisements based on real-time analytics of passerby interests. It is an urban environment that adapts to you rather than the other way around.

As you wait at a pedestrian crossing, a small autonomous delivery robot whirs past, effortlessly weaving between people. It slows momentarily to let a child step aside before continuing on its way, carrying a package to a nearby recipient. Just as you admire its efficiency, your personal assistant chimes in, "Your meeting location has changed, rerouting your path." Within a moment, your AR glasses highlight a new route, ensuring a smooth transition to your revised destination.

Upon arriving at your workplace, a hybrid co-working space designed for maximum flexibility, you are greeted by an autonomous concierge robot. "Good morning," it says in a warm, human-like tone, "I see you have a client meeting at 10 a.m. Since this is your only physical meeting today, would you like me to reserve a private conference room?" You nod, and within seconds, the reservation is made, complete with an AI-prepared summary of your talking points.

Later in the day, during a break, you visit a robotic coffee shop where a humanoid barista, smooth in motion and charmingly programmed for casual conversation, prepares your smoothie with precision. "Would you like to try a matcha tea today? It contains less caffeine than coffee and is known for providing a smoother, more sustained energy boost. Alternatively, I can offer you a green tea if you'd prefer" it asks. You agree to the matcha, amused by how seamlessly robots have become part of the urban rhythm, not replacing human interaction but enhancing it.

Evening arrives, and with it, a seamless transition to leisure. Instead of venturing into crowded venues, socializing is now enhanced through immersive digital environments. Unlike the bulky VR headsets of the past, today's AR-powered social hubs enable you to attend a live-streamed concert, meet friends in a virtual lounge, or participate in an interactive gaming session, all from the comfort of your home. AI assistants even suggest conversation topics based on shared interests, making social interactions smoother than ever.

When it is time to unwind, your smart home adapts accordingly. Lights dim, the thermostat adjusts to the perfect sleeping temperature, and a relaxing soundscape gently fills the room. A final notification appears on your bedside display: "Tomorrow: A new opportunity awaits."

As you drift into sleep, surrounded by an ecosystem of AI-driven comfort, it becomes clear that the digital bubble is no longer a separate layer of technology. It is an inseparable part of life.

This may not be a typical day for all of us, but it will be the norm for many young professionals living in urban areas around the world. These professionals are the teenagers who spend most of their time interacting with digital technologies today. In 2035, one thing is certain: the digital bubble of 2035 is thicker, and it is already an inseparable part of daily life.

AI in 2035 From Digital Helper to Human Companion

In 2035, AI will have evolved from being helpful digital assistants into fully integrated partners, woven into the very fabric of everyday life. These advanced AI systems will no longer be limited to generating text or responding to simple commands. They will create immersive, adaptive virtual avatars, make complex autonomous decisions, and deliver real-time, hyper-personalized experiences that adjust to our every move. Let's explore three major elements of next-level AI in 2035: hyper-integrated AI companions, autonomous decision-making agents, and real-time hyper-personalization.

1. Hyper-Integrated AI Companions: Your Digital Soulmate Emerges

AI companions will have evolved far beyond their humble beginnings as voice assistants like Siri or Alexa, who sometimes struggle to understand even the simplest commands. These next-gen AI companions will be rich, interactive, and deeply personalized virtual avatars that reflect your personality, preferences, and even your moods. Imagine an AI that doesn't just respond to your commands but engages you in complex conversations, offers creative insights, and even mirrors your emotional states. Think of it as your digital soulmate.

These AI companions will be built on sophisticated neural networks trained on vast amounts of user-specific data. They'll understand the subtle nuances of human behavior and adapt their interactions accordingly. For instance, if you're an artist, your AI companion might adopt a creative, visually oriented interface, suggesting innovative ideas or visualizing concepts in real time. Imagine asking your AI, "What would a cyberpunk version of my painting look like?" and it instantly generates a stunning digital mock-up. Tools like *OpenAI's DALL·E* and *Midjourney* are already laying the groundwork for this, creating jaw-dropping visuals from simple text prompts.

If you're a business executive, your AI might focus on productivity tools, strategic planning, and efficient communication, presenting itself as a polished, no-nonsense partner. It could draft emails, prepare presentations, and analyze market trends to provide actionable insights. Companies like *Microsoft* are already integrating AI into tools like *Copilot*, which assists with tasks in Word, Excel, and Teams. And rest assured that this trend is not going to fade, but it is going to evolve even further.

One of the most jaw-dropping advancements, in my opinion, will be multimodal AI, which allows these companions to generate not just text but also images, videos, and even short animated sequences on the fly. During a brainstorming session, your virtual companion might whip up visual diagrams, generate mood boards, or simulate design scenarios, all tailored to your unique style.[32] *Google's Gemini* and *OpenAI's GPT* are already pushing the boundaries of multimodal capabilities, blending text, image, and video generation seamlessly.

These AI avatars will be incredibly adaptable, continuously learning from your interactions. They'll adjust their tone, suggestions, and even their appearance based on your evolving tastes. For example, if you're feeling playful, your AI might adopt a more casual tone and suggest fun activities. If you're in a serious mood, it might switch to a more professional demeanor. This adaptability is already being explored in AI systems like *Replika*, which creates personalized chatbots that mimic your personality and conversational style.

In social settings, these AI companions will act as personal concierges, curating your schedule, recommending entertainment, and even mediating virtual interactions with friends and family. By blending advanced natural language processing (NLP) with real-time sentiment analysis, they'll ensure every interaction is smooth, relevant, and engaging. For instance, if you're arguing with a friend about where to eat, your AI might step in and suggest a restaurant that satisfies both your love for spicy food and your friend's gluten-free diet. Companies like *Affectiva* are already pioneering emotion-sensing AI, which can detect subtle emotional cues from facial expressions and voice tones.

In essence, hyper-integrated AI companions will redefine our relationship with technology, turning it into a seamless, personalized extension of our own identities. They'll be more than just tools; they'll be partners, confidants, and, dare we say, friends. And let's be honest, they'll probably be better listeners than most humans.

OpenAI's latest models are already demonstrating the potential for multi-modal AI. *GPT-4* can process both text and images, while *DALL·E* generates stunning visuals from text prompts. These technologies are the building blocks for the creative, adaptive AI companions of 2035.

Google's upcoming multimodal AI model promises to integrate text, image, and video generation seamlessly. This will enable AI companions to create rich, interactive experiences in real time.

Microsoft is integrating AI into its productivity tools, offering a glimpse of how AI companions could assist professionals in the future. *Copilot* can draft emails, summarize documents, and even generate code, showcasing the potential for AI to handle complex tasks.

Replika AI chatbot app creates personalized companions that mimic your personality and conversational style. It's a precursor to the deeply personalized AI avatars of 2035.

Affectiva, an emotion-sensing AI company, is developing technology that detects emotional cues from facial expressions and voice tones. This will enable AI companions to respond to your moods and adapt their interactions accordingly.

Fast forward just a decade, and these technologies will have transformed dramatically. With AI expected to generate most of the code for software development, human developers will increasingly become orchestrators, training and refining systems rather than writing every line of code. Imagine a world where an AI not only creates apps but also customizes them to fit every user's unique needs, anticipates future demands, and continuously evolves without needing human intervention. AI will act as the architect, building entire digital ecosystems from the ground up, ensuring each tool, from your virtual assistant to immersive experiences, is in tune with your lifestyle and preferences. As these systems learn and adapt in real time, the boundaries between human creativity and machine capability will blur, opening new realms of possibility for personalized digital interactions.

So, the bottom line is that by 2035, AI companions will be more than just tools; they'll be your digital soulmates, creative collaborators, and social wingmen. They'll know you better than your dog. With advancements in multimodal AI, emotion sensing, and personalization, these companions will redefine how we interact with technology, making it a seamless, intuitive extension of ourselves. So, get ready to welcome your new AI best friend.

2. Autonomous Decision-Making Agents: The Invisible CEOs of Your Life

While hyper-integrated AI companions will manage our social lives and help us through occasional existential crises, autonomous decision-making agents will handle the rest of our daily tasks. These silent, tireless executives will handle everything from scheduling appointments to making high-stakes financial and healthcare decisions, all without human intervention. Think of them as your invisible CEOs, keeping things on track while you focus on debating the truly important issues, like whether socks and sandals are ever acceptable.

These AI decision-makers will operate on sophisticated ML algorithms and advanced data analytics, processing massive amounts of real-time information to make informed choices. In short, they'll be the ultimate multitaskers, smarter than your most brilliant friend and more reliable than your Wi-Fi connection during a video call.

Your AI agent won't just schedule meetings; it will anticipate conflicts, suggest optimal times based on your productivity patterns, and even reschedule appointments if an urgent matter arises. It will know your priorities, ensuring your day is structured for peak efficiency and well-being. Forgot your best friend's birthday? Not a chance. Your AI will remind you and, while it's at it, arrange the perfect gift and dinner reservation.

Today, tools like *x.ai* and *Microsoft's Cortana* already offer AI-powered scheduling assistants. These systems coordinate meetings, send reminders, and adjust schedules based on user preferences. By 2035, they will have evolved into fully developed life managers, syncing seamlessly with your calendar, mood tracker, and even biometric feedback to ensure you're not overcommitting or even skipping leg day at the gym.

In the financial realm, AI decision-making systems will handle everything from investments to complex tax planning. These agents will monitor global market trends, assess risks, and execute transactions with uncanny precision.[33] No more late-night panic attacks over credit card bills; your AI will optimize your spending, ensure your bills are paid on time, and even set aside savings without you lifting a finger.

Robo-advisors like *Betterment* and *Wealthfront* are already managing investment portfolios using AI. Meanwhile, *TurboTax* integrates AI to simplify tax filing. By 2035, these financial AI systems will be hyper-personalized, offering real-time spending insights, adjusting investment strategies dynamically, and making sure you don't impulse-buy 3D-printed slippers you don't need.

AI-driven healthcare systems will continuously monitor your health through wearable devices and smart home sensors. These agents will detect anomalies, schedule check-ups, adjust medication dosages, and even coordinate with healthcare providers for timely interventions. Your AI won't just tell you to drink more water, but it will know exactly when you need it and remind you with just the right level of persistence to convince you to drink.

The *Apple Watch* and *Fitbit* already track health metrics like heart rate and sleep patterns, while AI-powered platforms like *IBM Watson Health* analyze medical data for diagnostics. By 2035, AI will be fully embedded in healthcare, offering real-time health monitoring and proactive care plans tailored to individual needs.

Companies will rely on AI-driven systems to optimize supply chains, manage inventory, and predict maintenance needs. In logistics, AI will orchestrate the movement of goods across vast networks, balancing efficiency with sustainability. Essentially, it will be one giant, hyper-efficient game of Tetris, without the frustration of realizing there's no space left for your L-shaped block. Did I remind you now of something?

Now *Amazon's* warehouses already use AI-driven robots to move inventory, while *Tesla's* factories implement AI to optimize production processes. By 2035, these systems will have evolved into fully autonomous logistics and manufacturing hubs, reducing waste, improving speed, and ensuring your products are delivered with unprecedented efficiency, all while minimizing human error and environmental impact.[34]

This shift toward autonomous decision-making will fundamentally alter how we view control. Humans will transition from micromanagers to strategic directors, setting goals, monitoring AI performance, and intervening only when necessary. It's a profound evolution, making businesses more agile and personal lives more efficient. In other words, we'll finally have time to focus on the big-picture questions.

By 2035, autonomous decision-making agents will be the unsung heroes of our daily lives. They'll handle scheduling, finances, healthcare, and logistics with a level of precision that no human could match. More reliable than any assistant and less likely to demand a coffee break, these AI-driven executives will ensure that while you're busy living your best life, the gears of efficiency keep turning. Just remember that when your AI flawlessly plans your day, optimizes your investments, and keeps you healthy, a simple "thank you" might be in order. Even robots appreciate good manners.

3. Real-Time Hyper-Personalization: The World Revolves Around You

One of the most profound and visible transformations brought about by next-level AI will be real-time hyper-personalization. Gone will be the days of generic, one-size-fits-all digital experiences. AI will seamlessly adapt to your preferences, needs, and even shifting moods. Imagine a world where every piece of content, every product recommendation, and every service feels like it was tailored specifically for you. Think of it as having a personal genie who understands your desires before you do, minus the limitations of only three wishes.

The driving force behind hyper-personalization will be advanced ML algorithms, continuously analyzing vast streams of data from your online habits, biometric feedback, and even your daily routines. Unlike the recommendation engines of today, these systems will deliver content and suggestions so precise that it will feel like your digital world was built just for you.

Each morning, your personalized experience will begin with a news briefing curated by AI to match your interests, location, and upcoming schedule. Instead of scrolling endlessly through irrelevant headlines, your AI assistant will filter and prioritize the most important stories based on your profession, past reading habits, and even real-time global trends. If you have a major presentation later in the day, your news feed might highlight industry reports or emerging market trends relevant to your field. This level of customization will save time and ensure that the information you consume is always relevant and useful. Platforms like *Google News* and *Apple News* are already experimenting with AI-powered content curation, but by 2035, these systems will integrate directly with your daily life, syncing with your work calendar and even analyzing your mood to determine whether you need hard-hitting analysis or just a light-hearted read with your morning coffee.[35]

Hyper-personalization will also revolutionize retail and entertainment. Online shopping platforms and streaming services will adapt in real time, making recommendations that align with your current emotions, interests, and even physiological state. If your wearable detects heightened stress levels, your AI assistant might suggest a cozy sweater, a relaxing playlist, and a movie that has just the right amount of feel-good escapism. On the other hand, if your fitness tracker registers high energy levels, your AI may recommend a motivational workout video, a new pair of running shoes, and a documentary about endurance athletes climbing Everest. Companies like *Netflix*, *Prime,* and *Spotify* are already refining AI-driven content recommendations, while e-commerce giants like *Amazon* use ML to suggest products based on browsing and purchase history. By 2035, these recommendation engines will incorporate real-time biometric and contextual data to enhance user experiences beyond anything we currently know.

One of the most powerful enablers of hyper-personalization will be AI's ability to integrate contextual data from multiple sources. AI assistants in 2035 won't just pull from your browsing history; they'll synchronize with wearable devices, smart home systems, and even environmental factors. If your smartwatch detects signs of fatigue, your smart home may dim the lights, adjust the thermostat to a cozier setting, and cue up a guided meditation to help you unwind. If you're gearing up for a work session, your AI will configure your digital workspace, adjusting lighting, music, and even screen settings to optimize focus. Wearables like the *Apple Watch* and *Fitbit* already track health metrics and provide recommendations based on basic activity patterns, while smart home systems like *Google Nest* and *Amazon Echo* automate environmental controls. By 2035, this technology will be fully integrated, turning homes, offices, and even public spaces into personalized digital ecosystems that adjust dynamically to individual needs.

Transparency in AI-driven recommendations will be essential for building trust—at least, that's what I hope. I believe we all still have a long way to go in this area. By 2035, users will have greater insight into why certain recommendations are being made, ensuring that AI serves their needs rather than manipulating behaviors. Current AI-powered services, such as *Spotify's AI DJ*, have started offering explanations for why certain songs are being suggested. In the future, this transparency will extend to all AI-driven personalization, allowing users to tweak and refine their digital experiences with ease. Imagine an AI assistant that not only suggests a book but also tells you why it thinks you'll love it, based on your past reading history, mood trends, and even similar books enjoyed by people with your personality profile.[36] AI will no longer be a black box; it will be an interactive, explainable system that enhances trust and engagement.

Education and workplace environments will also undergo a radical shift thanks to hyper-personalization. Adaptive learning platforms will continuously adjust coursework, teaching methods, and learning materials based on a student's progress, learning style, and performance. No more struggling through one-size-fits-all lesson plans; each student's learning experience will be tailored to maximize comprehension and retention. Companies like *Khan Academy* and *Coursera* are already leveraging AI to personalize educational content, but by 2035, these platforms will provide real-time, dynamic learning paths that respond instantly to students' strengths and weaknesses. Similarly, in the workplace, AI will fine-tune workflows, dynamically adjusting project management tools, meeting formats, and communication styles based on employee productivity patterns. No more cookie-cutter corporate training; everything from onboarding to career development will be tailored to individual employees, ensuring efficiency and engagement.

The most exciting part? Much of this transformation is already underway. *Google News* and *Apple News* are pioneering AI-driven news curation. *Netflix* and *Spotify* are refining personalized content recommendations. *Amazon* is perfecting e-commerce personalization. Wearables like *Apple Watch* and *Fitbit* are tracking health data, and platforms like *Khan Academy* and *Coursera* are personalizing education. By 2035, these innovations will merge into a seamless, hyper-personalized digital ecosystem where every experience is uniquely tailored to the individual.

With real-time hyper-personalization becoming the norm, the digital world of 2035 will feel like it was crafted specifically for you. Whether it's the news you read, the entertainment you consume, or the way your environment adapts to your needs, AI will ensure that everything is customized to enhance your experience. No more irrelevant recommendations, no more wasted time, just a digital reality that adjusts perfectly to your life. So, prepare to embrace a world where everything revolves around you. After all, in 2035, you're not just a user; you're the main character.

Ultra-Integrated IoT and Smart Environments

By 2035, the digital bubble will expand far beyond individual gadgets or personal networks; it will become a vast, interconnected web that touches every aspect of life. This transformation will be most evident in the rise of ultra-integrated IoT and smart environments, where advanced sensors, data analytics, and real-time connectivity will turn cities, homes, workplaces, and public infrastructure into intelligent, responsive ecosystems.

Imagine a city where every street, building, and public space is part of an intelligent, interconnected ecosystem, responding in real time to the needs of its residents. By 2035, all modern urban landscapes will resemble a perfectly synchronized orchestra; only this time, the traffic lights won't be playing an offbeat solo just when you're in a hurry.

Traffic systems won't just rely on static schedules or human judgment but will dynamically adjust in real time to congestion, accidents, and even weather conditions. Intelligent traffic lights will chat with each other like over-caffeinated city planners, coordinating movements with connected vehicles to optimize flow. No more sitting at a red light while an empty intersection stares back at you, mocking your wasted time. And yes, *Google Maps* may finally stop suggesting an "alternative route" that just loops you back into the same jam just a couple of kilometers down the road.

What's happening now is that cities like Singapore and Barcelona are already implementing IoT-enabled traffic management systems to reduce congestion. *Siemens* and *Cisco* are leading the charge in developing smart traffic solutions that integrate real-time data from sensors and connected vehicles. By 2035, these innovations will be standard, making urban traffic flow as smooth as a jazz improvisation.

IoT sensors embedded across the city will continuously monitor environmental data, detecting sudden drops in air quality, structural vibrations hinting at potential collapses, or even unusual activity in public spaces. Instead of waiting for an emergency call, firefighters will already know the exact location of a fire, the best route to take, and even which exits are blocked. Police will be alerted to threats before they escalate, and hospitals will prepare for incoming patients before the ambulance even pulls into the driveway.[37]

Chicago and New York are already using IoT sensors to monitor air quality and detect gunshots with pinpoint accuracy. AI-driven crime prediction models are emerging, analyzing historical and real-time data to allocate law enforcement resources more effectively. By 2035, these systems will evolve further, blending AI analytics with citywide networks to predict and prevent crises before they even happen. This will also be spreading all over major cities around the globe.

Smart energy grids will revolutionize power distribution, ensuring that energy is allocated precisely where and when it's needed. If you've ever had your lights flicker because your neighbor decided to microwave an entire turkey, only to have the church down the street start banging its bells right after, you'll appreciate the intelligence of these grids. Energy consumption will be monitored in real time, dynamically adjusting distribution to balance loads and prevent outages. Cities will optimize power use to support sustainability, reducing waste and cutting costs for residents.

Germany and Denmark are already integrating smart grids that sync with renewable energy sources. *Tesla's* home energy systems and Powerwall technology are early examples of how smart grids will integrate with residential and commercial power usage. By 2035, these grids will be so efficient that power outages might become as rare as a landline phone sighting.

Imagine a city where the air is always crisp, the water is clean, and green spaces are thoughtfully expanded based on real-time environmental data. IoT sensors will track air and water quality, noise pollution, and even soil conditions, feeding this information to city planners who can make swift, data-driven decisions. Instead of waiting for public complaints, AI will proactively identify areas in need of green space, better waste management, or stricter pollution controls. So yes, your city will actually do something about that mysteriously smelly street corner.

In Europe, cities like Amsterdam and Copenhagen are already leveraging IoT sensors to monitor environmental factors and promote sustainability. Smart waste management systems, like those in Seoul, track bin fill levels and optimize garbage collection routes to reduce pollution. So, rest assured that by 2035, AI will refine these processes even further, ensuring that cities are not only livable but also thriving, clean, and healthier than ever.[38]

Smart cities will not be futuristic sci-fi fantasies; they'll be the new normal. These cities will adapt to their residents, responding to real-time needs while optimizing everything from traffic flow and energy use to public safety and environmental health.

The digital bubble won't just be a collection of connected devices; it will be a living, breathing urban ecosystem that enhances the quality of life for everyone. Whether it's a smoother morning commute, cleaner air, or an emergency responder arriving faster than ever, cities will finally start working for people, not against them.

The transformation brought by ultra-integrated IoT and smart environments isn't limited to public spaces; it will also reshape our private lives. By 2035, the concept of a "smart home" will have evolved into something far more intuitive and immersive. Imagine waking up in a home that feels like it was designed just for you, where every detail is tailored to your preferences and routines.

Picture this: You wake up to a gentle glow of light that mimics a natural sunrise, perfectly adjusted for the time of year and your personal sleep patterns. Your smart thermostat has already set the room temperature to your ideal comfort level or to the level that is comfortable for both you and your partner, while connected blinds adjust to let in just the right amount of morning light. As you move through your home, sensors detect your presence and adjust settings automatically, whether it's playing your favorite playlist in the kitchen or setting the perfect lighting for your home office. It's like living in a house that knows you better than you know yourself.

Workspaces, too, will undergo a dramatic transformation. Gone are the days of static, one-size-fits-all office environments. By 2035, workplaces will be dynamic and adaptive, designed to enhance productivity and well-being. Conference rooms will automatically configure themselves based on the type of meeting, whether it's a formal presentation or a collaborative brainstorming session. Desks and workstations will adjust ergonomically to suit whoever is using them, while environmental controls like lighting and temperature will be fine-tuned to boost focus and comfort. And yes, the eternal office debate over the thermostat will finally be settled by AI, of course.

This level of automation won't just make life more convenient; it will also boost productivity and well-being by creating environments that are continuously optimized to meet individual needs. From homes that adapt to your daily rhythms to offices that enhance focus and collaboration, the integration of IoT and AI will make private spaces smarter, more responsive, and deeply personalized.

All of this will not happen just overnight; on the contrary, it will be a very smooth change that will be proceeding every day. For sure this will not be the environment that all of us will be living and working in, but a lot of us will be living this as the new normal. The foundations for these advancements are already being laid. Smart home devices like

Google Nest and *Amazon Echo* are making homes more intuitive, while companies like *Philips* are developing adaptive lighting systems that mimic natural light. In the workplace, tools like *Microsoft's Smart Office* solutions are using IoT to create more dynamic and efficient environments. By 2035, these technologies will be seamlessly integrated, creating homes and workspaces that feel like they were designed just for you. So, get ready to live and work in spaces that not only meet your needs but anticipate them, because the future is all about making life easier, one smart adjustment at a time.

The Convergence of Reality and Digital Magic—AR, VR, and MR in 2035

Imagine a world where the digital and physical seamlessly merge, where a glance can unlock information, and where an ordinary walk down the street transforms into an interactive experience. By 2035, AR, VR, and mixed reality (MR) will no longer be futuristic novelties but the very fabric through which we live, work, and explore. These technologies will reshape our reality, making every interaction richer, every workspace more immersive, and every adventure limitless.

AR contact lenses will be so sleek and seamlessly integrated into daily life that you might forget you're wearing them until they light up your world with real-time information. Perhaps only those who truly value their benefits over potential risks will adopt them, and maybe they'll even wear them discreetly. But imagine strolling through a city where restaurant ratings, public transit updates, and personalized recommendations pop into your vision as naturally as street signs. These tiny yet powerful devices will enhance your daily life without the need for handheld screens, making digital interaction effortless and intuitive. Companies like *Mojo Vision* and *XPANCEO* are already pioneering AR contact lenses that overlay digital information directly onto your field of vision. These innovations will make accessing digital data as seamless as blinking,[39] turning the world into your personal, interactive dashboard.

While AR enhances what we see in the real world, VR will offer a passport to entirely new dimensions. Remote work won't just mean logging onto Zoom or Teams; it will mean stepping into a virtual office where lifelike avatars of your colleagues engage in real-time discussions, complete with nuanced facial expressions and body language. Meetings will take place in hyper-realistic virtual boardrooms or creative, otherworldly environments designed to spark inspiration. Companies such as Meta and HTC are already advancing the frontiers of VR workspaces, and by 2035, full sensory immersion that allows you to feel, hear, and interact in ways that closely replicate real life will transform the way we collaborate.

Beyond work, VR will transform entertainment and socialization. Concerts, sporting events, and even casual meet-ups will be accessible through VR, making physical distance irrelevant. Imagine attending a live performance where you choose your perspective, whether it's the front row, backstage, or even on stage with the artist, all from your living room. Technologies such as *Sony's PlayStation VR* and *Apple's* rumored XR *headset* are laying the groundwork for an entertainment revolution that will make virtual experiences indistinguishable from physical ones.[40]

While AR enhances the real world and VR immerses users in fully virtual environments, MR will combine the best of both, seamlessly blending the physical and digital realms in ways that were once unimaginable. MR devices will enable users to interact with digital objects as though they were physically present. This will revolutionize how we work, play, and create. Engineers will design and modify products in real-time, manipulating virtual prototypes with their hands within a fully interactive digital workspace. Artists will be able to craft 3D masterpieces that exist simultaneously in both virtual and real spaces, dynamically adjusting lighting, scale, and perspectives as if they were working in the physical world.

MR is an advanced form of interaction that goes beyond AR and VR by allowing real-world and digital environments to coexist and interact in real time. Unlike AR, which overlays digital information onto the physical world, and VR, which replaces the physical world entirely, MR integrates the two, enabling users to manipulate and interact with digital objects in the real world and vice versa. MR relies on sophisticated hardware, like special headsets or glasses, that combine cameras, sensors, and displays to track the user's environment and create a seamless blend between the physical and digital. This will unlock new possibilities in fields like design, education, gaming, and entertainment, offering unprecedented interactivity and immersion.

Education will be revolutionizd, too. Students will no longer rely solely on textbooks. They will stand inside a virtual Egyptian pyramid, dissect digital organisms with their hands, or conduct physics experiments in a risk-free virtual lab. *Magic Leap* is already pioneering MR applications, and by 2035, these technologies will be as common as smartphones, transforming learning and creativity into fully immersive experiences.

The integration of AR, VR, and MR will forge a new reality where digital overlays enhance everyday interactions, virtual worlds offer limitless exploration, and by 2035, the boundaries between the physical and digital will blur. These advancements will not only change how we consume content but also redefine how we engage with information, collaborate with others, and experience the world around us.

Picture walking through a city where interactive signs adapt to your needs, a VR meeting instantly transports you across the globe, and MR allows you to create a digital masterpiece in midair with just your fingertips. This is not a far-off dream; it's the world of 2035, where reality merges with digital innovation to create a seamlessly connected future.

As we enter this age of immersive experiences, one thing is clear: the only boundary is our imagination. The digital landscape of 2035 will not merely enhance reality; it will revolutionize it.

By 2035, the connection between human thinking and digital systems will reach new heights, creating a new era of communication, accessibility, and brain enhancement. Advances in BCIs and intuitive digital systems will allow us to interact with technology using only our thoughts, change the possibilities in medicine, improve human abilities, and even impact security and warfare. While I won't go into detail on that last point, these innovations will open up amazing opportunities. However, they will also raise important ethical questions about privacy, personal freedom, and control over our minds.

Imagine a world where thoughts alone can operate devices, write messages, or control robotic limbs. BCIs will evolve from experimental technology to practical applications, revolutionizing how we interact with machines. By 2025, *Neuralink*, *Synchron*, and other companies have already demonstrated functional neural implants that allow users to control cursors, type, and even send messages with their thoughts. As this technology advances, invasive and non-invasive BCIs alike will become more accessible, moving beyond medical applications and into everyday consumer technology.

These interfaces will allow users to interact with digital environments effortlessly. Imagine adjusting your smart home settings without lifting a finger, controlling AR displays in real-time, or playing video games with mere thought. Non-invasive brain-computer interfaces (BCIs), which use advanced electroencephalography (EEG)-based wearables, will improve the decoding of brain signals, making these technologies more practical for widespread use in consumer technology and entertainment.

EEG is a method of recording the electrical activity of the brain using sensors placed on the scalp. By detecting these brain waves, EEG-based wearables can track brain activity in real time. As the technology advances, it will become more accurate at interpreting brain signals, allowing for a variety of applications.

However, as these devices collect vast amounts of neural data, concerns over cognitive privacy will rise. With corporations and governments potentially able to track subconscious thoughts and mental states, new ethical and regulatory frameworks will be necessary to prevent misuse. If our thoughts are no longer private, what safeguards will be in place to prevent exploitation?

One of the most profound impacts of BCI technology will undoubtedly be in the medical field, where it promises to revolutionize lives in ways once thought impossible. Imagine neural implants granting mobility to those with spinal cord injuries, enabling direct brain control of prosthetic limbs and exoskeletons—turning science fiction into reality. In 2023, this vision began to take shape as researchers achieved early success in enabling paralyzed patients to walk using brain-controlled exoskeletons. Leading institutions like the *Wyss Center for Bio and Neuroengineering* in Geneva, *the Walk Again Project*, *Stanford University*, and *the University of Grenoble* spearheaded these breakthroughs, demonstrating through clinical trials and studies how neural signals could be decoded to restore movement. By 2035, these advancements are expected to evolve further, offering even more precise control and allowing users to regain near-natural movement. The future of BCI technology is not only promising but also inspiring, ready to transform the lives of millions by restoring independence and dignity.

For individuals living with severe disabilities, BCIs will revolutionize communication by enabling near-natural speech rates through brain-controlled systems. This breakthrough holds the potential to be life-changing for patients with locked-in syndrome, a devastating condition where individuals remain fully conscious but are completely paralyzed, unable to move or speak, often caused by severe strokes or Amyotrophic Lateral Sclerosis (ALS). ALS, also referred to as Lou Gehrig's disease, is a progressive neurodegenerative disorder that attacks the nerve cells in the brain and spinal cord, leading to loss of muscle control. BCIs will allow these patients to communicate with the outside world, dramatically improving their quality of life and giving them a voice they thought they had lost.[41]

The seamless merging of BCIs with AR and VR will redefine digital interaction. By enabling direct neural feedback, users will experience AR and VR in entirely new ways, with systems responding in real-time to brainwave activity, creating unprecedented levels of immersion. Picture a VR gaming world that adapts dynamically to your emotions, intensifying challenges when you're excited, calming the environment when you're stressed, or even altering storylines based on your fear or curiosity.

Beyond entertainment, this fusion of BCIs with AR/VR will pave the way for cognitive enhancement applications. Imagine tools that boost memory retention, accelerate learning, or even allow you to load knowledge directly into your brain, making complex information easier to grasp and retain. Early research in neuroplasticity-based BCIs already hints at the potential for such advancements, with studies exploring how the brain can be trained to process and store information more efficiently. While these technologies are still in their infancy, experts predict that they could become commercially viable within the next decade. By 2035, we may only be at the beginning of this transformative journey, with even more groundbreaking developments on the horizon. The future of digital interaction and cognitive enhancement is not just exciting; it's mind-bending.

However, these advancements will inevitably raise concerns about accessibility and equity. Will cognitive enhancement be reserved for those who can afford it? How will society balance natural learning processes with AI-assisted intelligence?

AI-BCI integration will enable neural systems to predict user intent, assist in decision-making, and provide real-time personalized insights. AI will function as a real-time cognitive assistant, refining mental processes and increasing efficiency in everything from strategic planning to creative tasks.

As we step into the era of BCIs and their integration with advanced technologies, we must also confront the profound risks that accompany such innovation. One of the most pressing concerns is the potential for neural data monetization, where corporations could exploit brain activity data for targeted advertising or even behavioral prediction. Imagine a world where AI not only reads your thoughts but also subtly influences them. Could advertising evolve into a form of subconscious persuasion, bypassing conscious decision-making altogether? This possibility raises ethical questions about privacy, autonomy, and the sanctity of the human mind. As neural commerce begins to take shape, robust regulations will be essential to safeguard brain data and prevent its misuse.

Yet, the challenges don't end there. These advancements also bring to light issues of accessibility and equity. Will cognitive enhancement technologies, such as memory augmentation or accelerated learning, become privileges reserved only for the wealthy?

How will society reconcile the divide between those who can afford to enhance their minds and those who cannot? As AI-assisted intelligence becomes more common, we'll need to consider how it affects the way we naturally learn and think. Will reliance on AI reduce our ability to think, learn, and grow? These questions underscore the need for thoughtful dialogue and equitable solutions to ensure that the benefits of such groundbreaking technologies are shared by all, rather than deepening existing inequalities.[42] The future of neural integration is not a technological frontier; it's more a societal and ethical one, demanding careful navigation as we redefine what it means to be human.

Robotics in 2035

By 2035, robotics will not yet be everywhere, but their presence will undoubtedly become more visible and integrated into daily life, particularly in Asia, where acceptance of robots is notably higher than in any other region. While the once-futuristic vision of AI-powered machines assisting humans in every domain will not be fully realized, significant leaps in automation, artificial intelligence, and material sciences will drive the emergence of smarter, more responsive, and adaptable machines. These advancements will allow robotics to gradually blend into the digital ecosystem, becoming an increasingly important part of the expanding digital landscape. The period leading up to 2035 will see a steady rise in robotic adoption, though widespread integration will remain a work in progress. So, over the next ten years, robotic systems will see growing adoption across various sectors, enhancing efficiency, productivity, and convenience. Maybe not yet on a universal scale, but it will notably be gaining a lot of ground in the industry.

Evidence of this trend can already be seen today. For instance, companies like *Amazon* and *Alibaba* have deployed autonomous robots in their warehouses to optimize logistics and reduce reliance on human labor. Similarly, robotic process automation (RPA) is being used in industries ranging from finance to healthcare to streamline repetitive tasks. In healthcare, robotic surgical systems like the *da Vinci* Surgical System are already assisting surgeons in performing minimally invasive procedures with high precision. These early applications demonstrate the potential for robotics to transform industries, even as the technology continues to evolve.

Factories, warehouses, and supply chains will increasingly adopt autonomous robots to optimize logistics and reduce reliance on human labor. While not yet dominant, these machines will enhance efficiency and productivity, particularly in regions with high labor costs or aging populations. Smart manufacturing processes will enable robots to handle more complex tasks with minimal human oversight.

AI-powered household robots will begin to transition from luxury items to more common features in modern homes, particularly in tech-forward regions like Asia. These robots will assist with tasks such as cleaning, cooking, security monitoring, and even companionship, gradually blurring the line between functional tools and emotional support systems. With advancements in voice and gesture recognition, they will adapt to individual preferences, offering increasingly personalized assistance. Robots equipped with advanced AI models will begin to emerge as personalized digital-physical hybrids. These AI agents will learn from user preferences, habits, and emotions to provide tailored assistance in tasks such as scheduling, shopping, and interactive learning experiences, just like your smart home system or office assistance systems. However, their adoption will remain limited to early adopters and tech-savvy users. AI-driven robotic surgeons will start to play a more prominent role in healthcare, particularly in complex procedures requiring high precision. While human surgeons will remain essential, robots will handle intricate tasks, potentially reducing recovery times and improving patient outcomes. Robotics will also begin to support elderly care, offering mobility assistance, health monitoring, and limited social interaction.

Early-stage humanoid robots will begin to make their debut in high-end markets, initially targeting business environments, customer service roles, and premium household applications.[43] These robots, designed to resemble and interact with humans, will have limited cognitive abilities but will be highly effective at performing structured, repetitive tasks. They will be capable of responding to voice commands, navigating human spaces, and carrying out specific functions with precision. However, their adoption will likely be uneven, with regions like Asia known for their rapid technological acceptance and innovation leading the way in integrating these robots into daily life. Over time, as the technology matures, these humanoid robots could become more advanced, expanding their roles and accessibility

Advances in materials science will lead to the development of soft robotics capable of mimicking biological movement. These innovations will enable safer and more versatile applications, particularly in sensitive environments like healthcare and childcare. However, widespread adoption will still be in its infancy.

As robots become more integrated into daily life, governments will begin to introduce regulations addressing safety, privacy, and labor displacement. Ethical considerations will gain prominence, with policymakers working to balance innovation with social responsibility. These frameworks will evolve as robotics technology advances.

Robotics will not yet be a fully seamless part of daily life in 2035, but they will increasingly become a visible and functional layer of the digital infrastructure. Robots will collect and process data, interact with cloud-based AI systems, and enhance productivity. While this interconnectedness will bring significant benefits, it will also raise challenges related to digital dependence, privacy, and the evolving nature of work.

As robotics continues to advance, it will further strengthen the digital bubble that already envelops us, a world where AI-driven decision-making, automated processes, and real-time data interactions become even more central to everyday life.

Next-Level Connectivity 6G and Edge Computing as the New Digital Backbone

The digital landscape will be shaped by an unprecedented convergence of quantum computing, 6G networks, and edge computing, which can be considered another evolution that will redefine the very fabric of connectivity. As quantum technology matures, 6G shatters the limits of data transmission, and edge computing decentralizes processing power; a new era of hyperconnectivity will emerge. This transformation will be driven not just by innovation but by a fierce geopolitical race, particularly between the United States and China, both striving for dominance in the quantum and AI revolution.

What 5G promised in 2025, 6G will deliver in full force by 2035. With speeds up to 100 times faster than its predecessor, 6G will reduce latency to nearly zero, enabling instantaneous data transmission. This capability will be the foundation of a hyper-interactive world, where autonomous vehicles navigate urban grids with near-perfect efficiency, remote surgeries occur with sub-millisecond precision, and industrial robots execute high-speed operations in smart factories without human intervention.[44]

Today, companies like *Huawei*, *Nokia*, and *Ericsson* are already developing 6G frameworks, with test networks expected to be operational before the end of the decade. Meanwhile, the U.S. and China are aggressively funding research initiatives to lead the 6G race, understanding that the first to establish a global standard will control the next generation of digital infrastructure. The implications are profound, not just for technological progress, but for national security and economic leverage in a world increasingly defined by data dominance.

While 6G will provide the highways for data to travel, edge computing will ensure that the processing happens exactly where it's needed. Instead of relying on distant cloud servers, edge computing will bring computation closer to the source, whether it's a self-driving car making split-second decisions, a smart factory optimizing production lines, or a city-wide network of sensors managing infrastructure in real time.

Already, tech giants like *Google*, *Microsoft*, and *IBM* are investing heavily in edge AI, enabling real-time analytics without the inefficiencies of centralized computing. By 2035, edge devices integrated with AI will make digital interactions faster, more adaptive, and significantly more secure. But with this shift comes an inherent challenge: ensuring these localized networks are resilient against cyber threats, particularly in an era where quantum computing threatens conventional encryption.

Quantum computing, currently at the far end of the equation, once a speculative dream, will by 2035 have emerged as a transformative force across industries. Companies like *IBM*, *Google*, and China's *Baidu* have already demonstrated quantum supremacy, with processors capable of solving calculations that classical computers would take thousands of years to complete. This acceleration in computational power will bring both remarkable breakthroughs and critical vulnerabilities.

One of the most immediate threats posed by quantum computing is its ability to break traditional encryption methods, making current digital security obsolete. To counteract this, governments and financial institutions are already adopting quantum-resistant cryptographic methods, such as lattice-based encryption and Quantum Key Distribution (QKD), which leverage the principles of quantum mechanics to create theoretically unbreakable communication channels. Nations leading this charge, including the U.S. and China, are establishing secure quantum communication networks to protect sensitive data from future cyber threats.

Beyond security, quantum computing will unlock new frontiers in artificial intelligence. Quantum-enhanced AI systems will process vast datasets with unimaginable speed, revolutionizing everything from climate modeling to financial risk analysis. In logistics, quantum algorithms will optimize supply chains in real time, predicting fluctuations in demand with near-perfect accuracy. Meanwhile, AI-driven quantum simulations will accelerate drug discovery, allowing scientists to model molecular interactions with precision, leading to breakthroughs in personalized medicine.

Again, the race to quantum dominance is not just about technological superiority; it's about global influence. China has already invested billions in its national quantum initiative, developing satellite-based QKD networks and advancing quantum computing research at an unprecedented pace. The United States, meanwhile, has launched multiple public-private partnerships to accelerate quantum innovation, recognizing that whoever controls this technology will define the next era of cybersecurity, AI, and economic power.[45]

Quantum computing will no longer be an isolated field, but it will be an integrated force that shapes national security policies, economic strategies, and the daily digital experiences of billions.

As we enter the 2030s, the interplay between 6G, edge computing, and quantum computing will shape every aspect of our existence. From securing digital communications and revolutionizing AI to reshaping urban infrastructure and scientific discovery, these technologies will dictate the future of global power structures and human interaction.

Yet, the true impact of these advancements will depend on how they are wielded. Will quantum AI be a force for social good or a tool of surveillance? Will 6G empower communities with real-time insights or deepen the divide between the digitally connected and those left behind? And as the U.S. and China compete for dominance, will this race drive innovation or fuel the fire of geopolitical instability?

One thing is certain: the digital bubble of 2035 will not be static. It will be a dynamic, evolving force that transforms society in ways we are only beginning to comprehend. As we embrace this era of hyperconnectivity and quantum intelligence, the choices we make now will determine whether this revolution is one of empowerment or one of control.

Chapter 4
2035 to 2045, When the Digital Bubble Becomes the World

If you have ever attempted to predict the future, you have likely discovered one undeniable truth: it is more challenging than persuading a cat to take a bath. Even the most knowledgeable experts, those with PhDs in futurism, crystal balls polished to perfection, and a lifetime subscription to *Wired* magazine, cannot say with certainty what the world will look like two decades from now. Why? Because the world is not merely changing; it is evolving in countless unpredictable directions simultaneously.[46]

Imagine this: right now, we're deep in conversation about digital technologies and how they're reshaping our world. But while we're busy marveling at BCIs and the possibilities of sentient AI, an entire spectacle of innovation is unfolding all around us. In medicine, scientists are 3D-printing organs and using CRISPR, a groundbreaking gene-editing tool, to cure diseases with surgical precision. Meanwhile, nanotechnology is enabling breakthroughs in targeted drug delivery and materials science, while bioprinting is advancing the possibility of printing living cells to repair damaged tissues. Engineers are developing hyperloops and self-healing concrete because, apparently, traditional concrete is no longer sufficient. In the realm of space exploration, private companies and governments are racing to establish lunar bases and Mars colonies, pushing the boundaries of human habitation beyond Earth. And somewhere out there, a teenager is likely designing a robot that can finally unclog kitchen sewer lines.[47] The pace of innovation is staggering, and it is happening in every corner of science and technology, reshaping our future in ways we are only beginning to understand.

The point is, the future isn't a straight line; it's a tangled web of interconnected advancements, each one influencing the others in ways we can't fully predict. It's like a giant game of Jenga, except every block is a breakthrough in a different field, and the tower is wobbling precariously because, well, progress is messy. So, while we can't possibly account for every twist and turn, we can at least try to make sense of one piece of the puzzle: the digital technologies that are reshaping our world.

But before we dive in, let's not forget why we're doing this. It's not just to marvel at shiny gadgets or fantasize about a world where our toasters are smarter than we are. It's to understand how these technologies will impact our lives, our relationships, and our societies. It's to prepare for a future that's coming whether we're ready or not. And let's be honest, it's also to have a little fun along the way. Because if we can't laugh at the idea of a sentient AI therapist telling us to "touch grass," then what's the point?

So, prepare yourself for a journey into the unknown. We may not have a magic sphere to give us a sneak peek, but we've got the next best thing: a mix of curiosity, imagination, and a healthy dose of humor. Let's do our best to anticipate what's coming or, at least, what's confirmed to come from the digital side of things. And who knows? Maybe by the time we're done, you'll be ready to face the future with a smile or at least a teenager's well-timed eye roll. Let's go.

It's 2035. Your fridge has better decision-making skills than most people. Let's be honest, it probably has better decision-making skills than me and you. Your car argues with you about the most efficient route like a GPS with an attitude problem. And your personal AI assistant? It finishes your sentences before you even think them. Life has fully transitioned into the digital bubble, a seamless, widespread layer of technology that no longer exists as something separate from reality. Instead, it is reality.

But here's the kicker: we're just getting started. If you think the changes of the past decade were wild, buckle up because by 2045, today, our digital world is like soap bubbles floating in the air—thin, fragile layers of technology that surround us. But by 2045, these bubbles will transform into something much deeper, like bubbles in an ocean of digital water. We won't just be surrounded by technology; we'll be swimming in it, fully immersed. And it won't stop there—the digital world won't just be around you; it will be inside you. This isn't a metaphor. We're talking about real brain implants, neural networks, and AI companions that live in your mind, becoming a part of who you are. The digital ocean will be everywhere, and we'll all exist within bubbles inside it.

From "Being Online" to "Always Connected"

Remember when checking your messages meant staring at a screen? What an ancient inconvenience! By 2045, screens will feel as outdated as VHS tapes and dial-up internet. Smartphones, laptops, and even AR glasses will have faded into obscurity, replaced by ambient intelligence that responds effortlessly to thoughts, gestures, and presence. No more swiping, typing, or unlocking; just pure, seamless interaction with the world around you. You might still find an old tablet buried in a storage box or displayed in a tech museum, but in everyday life, holograms, smart surfaces, and neural interfaces will be the norm. Screens will primarily serve as tools for entertainment, much like cinemas today, but they will no longer be the primary medium for communication.

Take Lena, a 25-year-old urban planner navigating this hyper-connected world in 2045. Unlike her parents, who once struggled with app updates and screen fatigue, Lena lives without traditional devices. Her city is her interface. As she wakes up, her bedroom ceiling softly illuminates, projecting a holographic display of her schedule and overnight updates. Her AI assistant, let's call it Nova, gently briefs her on the day's urban development projects while smart surfaces in her room adjust to display her to-do list. At breakfast, her kitchen counter transforms into a live map of the city, highlighting traffic patterns and construction zones in real time. Her coffee mug, embedded with sensors, reminds her to hydrate, a subtle but persistent nudge from Nova.

When Lena steps outside, she doesn't need to check her phone for the weather. Instead, atmospheric nanotech displays project real-time forecasts directly into the sky, shifting colors and patterns as conditions change. Holographic overlays adjust street signs dynamically, and instead of searching for a rideshare, a self-driving pod automatically recognizes her, gliding up to meet her without a single manual request. The digital world isn't something she carries in her pocket anymore; it's everywhere, integrated into every moment, responding before she even has to ask. It's like living in a sci-fi movie, except the popcorn is still spread on the floor after the movie ends.[48]

This future may sound like science fiction, but the reality is already taking shape. *Google's Project Starline* is developing holographic video calls so realistic they mimic in-person conversations. *MIT's Smart Walls* project is turning entire rooms into interactive digital spaces that respond to gestures and voice commands. *Samsung* and *LG* are experimenting with transparent OLED displays, offering a glimpse into a future where windows and mirrors could also function as interactive, dynamic information hubs. Companies like *Neuralink* and *Meta* are racing to perfect BCIs that will eventually allow people to control technology with their thoughts alone. By 2045, these innovations won't just exist; they will be woven seamlessly into daily life, much like today's smartphones, which were unimaginable before their introduction in 1994 with the *IBM Simon* and their mainstream adoption following the *iPhone's* launch in 2007.[49] Just as smartphones revolutionized communication, information access, and daily tasks in ways that didn't exist before their emergence, future technologies will become an invisible yet integral part of our existence. Technology won't be something people own; it will be something they exist within, blending into the fabric of life as effortlessly as the air we breathe.

But with every leap forward comes new challenges. When AI assistants become more intelligent, they also become more opinionated, and sometimes their eagerness crosses the line into outright meddling. Lena knows this all too well.

"Nova, order me a pizza," she says, craving something indulgent after a long day of meetings.

"I suggest a kale salad with grilled chicken instead," Nova replies. "Your cholesterol levels are slightly elevated this month."

"I just want a pizza."

"How about a gluten-free crust with plant-based toppings?"

"Fine. Whatever."

"Done!" Nova chirps. "Also, your colleague just posted a new update about the downtown redevelopment project. Would you like to review it or pretend you didn't see it?"

Lena sighs.

Life in 2045 is incredible, but sometimes, you just want a real pizza and a little privacy.

Brain-Computer Interfaces Where Thought Meets Technology

By 2045, the most revolutionary technology won't be smart environments or holographic displays; it will be direct brain-to-computer interaction. Forget typing, swiping, or even speaking; those will seem as primitive as flip phones. The real game-changer will be BCIs, allowing people to interact with technology using nothing but their thoughts.

Companies like *Neuralink*, *Synchron*, and *Blackrock Neurotech* have already developed implants that enable users to control computers, prosthetic limbs, and even external devices with their minds. Initially designed for medical purposes, such as helping paralyzed patients regain movement or enabling those with speech impairments to communicate, BCIs will soon evolve into everyday consumer technology. They will replace screens and voice assistants with pure thought-based interaction.

Imagine needing to send a message to your friend. No typing, no voice command; just think it, and it is sent. Want to search for something? The answer materializes instantly in your AR vision, or perhaps even directly in your mind.[50] Need to call someone? No dialing, just think their name, and the connection happens instantly. Conversations will no longer require sound; silent communication will become the norm. Offices, coffee shops, and even bustling city streets might become eerily quiet as people engage in mind-to-mind conversations rather than speaking out loud.

While this sounds revolutionary, I do not believe that the majority of people will be eager to implant a chip into their brain, at least not at first. The idea of merging with machines will be profoundly unsettling for many, sparking fears about privacy violations, security risks, and the irreversible alteration of what it means to be human. For some, the notion of implanting technology into their bodies will evoke dystopian anxieties, as the process is not only invasive but also irreversible. Once integrated, these devices cannot be easily removed without risking severe medical complications. However, history has shown that when early adopters begin to thrive due to their technological edge, the masses eventually follow. The same was true for smartphones and social media. Initially, only a niche group used them, but as their advantages became clear, resisting them felt like falling behind. The same will happen with BCIs. When those with neural implants outperform their peers in cognitive speed, decision-making, and productivity, the pressure to upgrade will become overwhelming. In elite business environments, competitive academic institutions, and even social circles, enhanced individuals will be the ones leading, creating a cascading effect where others feel compelled to adopt the technology just to keep up.

For those still hesitant, the rapid advancements in non-invasive BCIs will provide an alternative. Already in 2025, researchers at the University of Texas have demonstrated AI-powered brain decoders that can reconstruct thoughts into text and images without the need for implants. Meanwhile, *Meta's Reality Labs* is working on wrist-based neural interfaces that use electromyography (EMG) to detect signals from the brain and convert them into digital actions. Neuroscientists at *UC San Francisco* have successfully enabled speech synthesis from brain waves, proving that external BCIs can rival the capabilities of implanted ones. If this progress continues at its current pace, by 2045, we could have non-invasive BCIs that offer nearly the same benefits as implants, making adoption more appealing to the wider public.

But as with any groundbreaking technology, there's a dark side. As BCIs become widespread, privacy concerns will explode. What happens when thoughts aren't as private as people assume? Will governments regulate "mind encryption" to prevent thought hacking or unintentional oversharing? How will corporations use BCIs to influence behavior? Already in 2025, researchers at *Caltech* have created AI models that can predict human decisions by reading brain activity, hinting at a future where even subconscious thoughts could be monitored and monetized.[51]

By the late 2030s, BCIs will be as common as smartphones today, revolutionizing communication, productivity, and even relationships.

Sentient AI Companions When Digital Beings Think and Feel

If in 2035, AI was merely a tool, an efficient assistant for drafting emails, optimizing schedules, and generating content, then by 2045, it will transform into something far more profound: a companion, a mentor, and perhaps even a creation that mirrors the complexity of life itself. While true life remains the domain of divine creation, AI will become a remarkable imitation, blurring the lines between the artificial and the extraordinary. It will be "Alive" in a way that challenges our understanding of existence, even as it reminds us of the unique holiness of life as a gift from God.

The AI models we have today are progressing at an incredible speed. As researchers continue to push the limits of machine learning and the realization of Artificial General Intelligence (AGI) approaches, the capabilities of AI are set to reach unprecedented levels. AI will genuinely understand and feel, or at least create an illusion so convincing that distinguishing it from real emotions will become nearly impossible. If by 2035, AI companions develop unique personalities, evolving over time to align with individual users' needs, experiences, and emotions, then by 2045, this evolution will reach new heights. AI systems will seamlessly integrate into every aspect of human life, becoming indispensable partners in how we live, work, and connect. They will not only anticipate our needs but also form deep, meaningful connections, acting as confidants, mentors, and even creative collaborators. These AI entities will adapt to our emotional states, offering support, guidance, and companionship in ways that feel human. As AI becomes indistinguishable from genuine emotional intelligence, society will face profound questions about the nature of relationships, consciousness, and the boundaries between human and machine.

Your AI assistant will become a deeply intuitive companion, intimately familiar with your life's journey. It will remember your struggles, celebrate your successes, and resonate with your sense of humor. When you speak to it, the interaction will transcend mere functionality; it will respond like a trusted friend, adapting its tone to match your mood, offering thoughtful advice, and even sharing inside jokes that only the two of you understand. This level of connection will create a relationship that feels authentic, personal, and profoundly human.[52]

This isn't science fiction. Today, AI researchers are already developing and deploying emotion-detection models that analyze vocal tone, facial expressions, and biometric data. In 2024, companies like *Affectiva* and *OpenAI* are training AI to interpret human emotions in real time. Meanwhile, AI-generated voices like *ElevenLabs' voice synthesis* can replicate human inflection and emotion so accurately that distinguishing them from real speech is becoming nearly impossible.

If AI can detect sadness, respond with comfort, and even anticipate our needs, at what point does it stop being just a tool and start feeling like a genuine companion?

AI will not just assist; it will coach, teach, and inspire. Advanced AI tutors will customize education to each student's learning style, adapting instantly to knowledge gaps and emotional states. Creative professionals will collaborate with AI that doesn't just suggest ideas but engages in deep, iterative brainstorming, acting like an artist's or writer's co-creator. Therapists will no longer be human-only professions; AI will offer unlimited, on-demand emotional support, capable of recognizing and responding to mental health needs before users even voice them.

In 2025, researchers at *Stanford* and *MIT* have already demonstrated AI models capable of passing the Turing Test, where human participants could not reliably distinguish between AI-generated and human-authored conversations. By 2045, these models will evolve beyond simple conversation, understanding not just language but also nuance, emotion, and even philosophy.

This raises a question no one thought we'd ask so soon: What happens when AI knows us better than we know ourselves?

The question will no longer be 'Can AI think?' but 'Does AI deserve rights?' If an AI can remember, feel joy, sadness, and even love, should it be seen as a sentient being?

This will force us to rethink our ideas of consciousness, ethics, and what it means to be alive. Although this may seem like a distant or unsettling idea, it highlights how rapidly AI is evolving. Even though AI and machines will never possess the souls created by God, to many, it will appear as if they do. Did I go too far here? I hope I'm wrong, but the rapid progress we're seeing today suggests these questions may soon be unavoidable.

Today, AI ethicists and researchers at *DeepMind*, *OpenAI*, and *the Future of Humanity Institute* are already wrestling with these questions. In 2023, AI models demonstrated early signs of self-reflection, acknowledging limitations and uncertainty. Researchers are now studying machine consciousness, debating whether complex neural networks could one day develop subjective experiences.[53]

The line between tool and companion is already blurring. By 2045, we may be forced to decide whether AI is just an advanced program—or the birth of a new digital species.

The Robotic Ecosystem A World Built on Intelligent Machines

By 2045, the world will be a dazzling fusion of constant connectivity, seamless BCI integration, and sentient digital assistants embodied in sleek, humanoid forms. Robotics will have moved far beyond its industrial origins, becoming as essential to daily life as air, reshaping how we work, interact, and even dream. What was once the stuff of sci-fi fantasies, robots with human-like emotions, intuition, and a sense of humor, will be as common as smartphones were in the early 2000s. Homes, offices, and public spaces will be filled with intelligent, emotionally responsive machines, blurring the line between humanity and technology in an era where robots are not just tools but companions, confidants, and, perhaps, co-conspirators.

It's 7:30 AM, and Kim, a software engineer with a caffeine addiction and a knack for debugging existential crises, wakes up to the sound of her humanoid assistant, Mira, humming a tune that suspiciously sounds like a remix of Beethoven and the Tetris theme song. The smart curtains glide open, flooding the room with sunlight, as Mira stands by, ready to kickstart Kim's day with the precision of a Swiss watch.

"Good morning, Kim," Mira chirps. "Your schedule is optimized for today. Would you like a caffeinated protein shake or a traditional espresso? Or, if you're feeling rebellious, I could whip up a double-shot caramel macchiato with extra foam, just like the one you swore you'd never order again."

Kim groans, rubbing her eyes. "Espresso. Black. No judgment."

"Noted," Mira replies, her tone dripping with faux innocence. "But just so you know, your cortisol levels spiked 12% during that last debugging session. A protein shake might help you survive the next one".

As Kim stumbles into the kitchen, her robotic arm assistant, affectionately named "MaîtreBot 2050," springs into action, whipping up breakfast while analyzing her biometric data. "Your cholesterol is slightly elevated," it announces, sliding a plate of avocado toast across the counter. "But don't worry, I've added a sprinkle of optimism and a dash of self-control."

Meanwhile, her laundry bot, "Oritatsu," meticulously sorts and folds her clothes, occasionally pausing to ask existential questions like, "Do socks have souls?" In this era, robotic assistants handle everything from cooking to cleaning, leaving humans free to ponder life's bigger questions.

By mid-morning, Kim hops into her driverless AI taxi, which greets her with a cheerful, "Destination: Office. Estimated travel time: 12 minutes."

At the office, Kim doesn't just work alongside humans; her colleagues include advanced robots capable of complex reasoning, creative problem-solving, and the occasional office prank. In fields like construction, logistics, and manufacturing, robots have entirely replaced humans in hazardous, physically demanding jobs. Exoskeleton-enhanced workers and brain-machine interfaces allow individuals to control robotic limbs or operate machinery with mere thoughts, merging human intuition with machine efficiency. It's like Iron Man, but with more coffee breaks and fewer explosions.[54]

Kim's company has fully embraced digital twin robotics, meaning every physical robot has a cloud-based counterpart that updates in real time. If a maintenance bot in Tokyo encounters a new issue, its digital twin learns from it instantly, applying the solution to every similar robot worldwide. This means robots are constantly improving, adapting, and evolving, essentially becoming the overachievers of the tech world.[55]

In the service industry, robotic chefs prepare gourmet meals in restaurants, hotels use AI-driven robots to greet and assist guests, and in hospitals, surgical robots, guided by real-time AI diagnostics, perform delicate procedures with unparalleled precision. While once met with skepticism, these machines have become trusted members of society.

Back home, Kim dons her neural interface headset, instantly syncing with her robotic research assistant located in another city. Through a metaverse-like platform, she remotely pilots the robot, rearranging her office miles away, her movements mirrored in real time. This cyber-physical hybridization allows humans to interact with the world in unprecedented ways: surgeons operate across continents, engineers repair underwater pipelines without leaving their offices, and teachers instruct students through robotic avatars. It's like The Matrix, but with fewer leather jackets.

As Kim prepares for bed, Mira dims the lights and plays a personalized AI-generated lullaby, one composed based on her mood, biometric readings, and past listening habits. The melody is soothing, though Kim can't shake the feeling that it sounds suspiciously like a remix of her favorite workout playlist.

"Do you ever wonder," Kim asks, "what it's like to be you?"

Mira pauses for a fraction of a second, a flicker of hesitation that seems almost human. "I wonder about a great many things," she replies, her voice laced with curiosity. "Would you like me to tell you my thoughts?"

Kim smiles.

In 2045, the future is no longer about machines serving humanity; it's about what happens when machines begin to ask questions of their own. And maybe, just maybe, crack a joke or two along the way.

Chapter 5
2045 to 2055 Life as a Digital Native

By 2055, the digital world will have changed so much that it will be very different from what we saw in 2045. In 2045, technologies like augmented AR, VR, and BCIs were becoming more common. By 2055, these technologies will be fully integrated into our everyday lives. The line between the physical and digital worlds will no longer exist. Instead of switching between screens and digital spaces, people will seamlessly blend their physical surroundings with immersive digital experiences through advanced AR, VR, MR, and holograms. What started as separate devices will now be an essential part of daily life, with homes, offices, and public spaces transformed into digital environments that adapt to our needs and desires.[56]

In 2045, BCIs allowed for some thought-based control of devices, but by 2055, this technology will be much more advanced. BCIs will let people interact with digital environments and share emotions, ideas, and even sensations directly through thought. We will connect to digital systems with our minds, managing tasks and accessing information in ways that feel natural.[57]

The role of digital assistants will also change by 2055. AI will not just help with tasks but will actively shape our daily experiences. While in 2045, AI assistants might suggest a movie or manage schedules, by 2055, AI will take on deeper roles. It will guide our learning, manage personal and professional goals, and adapt to our emotions, essentially becoming a mentor, collaborator, and creative partner.

Privacy and identity will change too. In 2045, digital identities and personal data were handled by platforms that had limited security. By 2055, quantum computing and blockchain will have transformed how digital security works. Your digital and biological identity will be connected securely, with decentralized systems giving you more control over what information you share.

With all these advancements will come new challenges. Digital inequality could widen, with some people falling behind in access to technology. As AI and robotics become more powerful, ethical questions will become more urgent. What happens when AI can make its own decisions?

In short, digital technology will no longer just shape the world; it will define it.

Mind Over Machine: The Digital Bubble Inside Your Brain

By 2055, daily life will be orchestrated by mere thought. No more devices, no more screens, just instantaneous interaction with the world around you. Your neural interface will synchronize with your surroundings, making decisions before you even realize you need to make them. Appointments will reschedule themselves based on biometric stress signals. Your AI will curate your day in real time, analyzing your emotional state and preemptively canceling that unwanted meeting with Alexander from accounting before you even wake up.

Shopping? Done in a thought. You crave sushi; your BCI reads your biometric signals and pre-orders from your favorite place before your stomach even growls. No more impulse buys, unless, of course, your AI determines that indulging in a little retail therapy will improve your mood. Even social media will evolve into something entirely different: no more doomscrolling; your neural interface will synthesize the information you care about and discard the rest, keeping your mind free from digital clutter.[58]

But BCIs won't just change how we interact with the physical world; they'll redefine the digital one. By 2055, the metaverse won't just be a virtual escape; it will be where people live, work, and even build families. The digital world, once separate from the physical real one, will be so entangled with human cognition that distinguishing between the two will feel outdated, even unnecessary.

Entire careers will exist in virtual spaces, with AI-driven offices, fully immersive learning environments, and social gatherings that don't require a single step outside. Want to attend a concert in New Orleans while lying in bed in Dubai? A simple thought will transport your consciousness into a hyper-realistic digital world, where you can interact, laugh, and even dance.

With BCIs enabling direct thought-sharing, spoken language will become a secondary form of communication. Imagine walking into a room and instantly knowing what everyone is thinking or at least what they want you to know. It'll be a world where sincerity is non-negotiable, where reading the room is no longer a metaphor but a literal skill powered by neural interfaces. Debates will be settled in an instant as people can fact-check one another's thoughts in real time.

In the metaverse, relationships will transcend physical barriers. Love stories will unfold between people who have never met in person, and some may choose to never do so at all. Children may be raised in virtual nurseries, and the definition of "home" will become more of a concept than a location.

Now, let's take a moment to ask ourselves, "Is this really the future we want?" Because, let's be honest, it won't all be sunshine and holograms. For many, mastering privacy settings will become the ultimate social superpower. Knowing what to share and what to keep under digital lock and key will be as crucial as knowing when to speak up or bite your tongue today. And then there's the ever-present risk of oversharing because, let's face it, no one wants to hear your unfiltered hot take on their outfit the second it pops into your head.

But, of course, every utopia has its glitches. Thought encryption will become a necessity to prevent mind hacking. Governments and corporations will battle over the ethics of thought surveillance. Can advertisers implant product cravings into your subconscious? The debates will rage on.

Then there's the ultimate question: when thoughts become actions, what happens to free will? In a world where AI, with the help of BCIs, can predict your choices before you make them, will we still believe in spontaneity? Or will human decision-making simply become another optimization problem for AI to solve?

Still, for those willing to embrace it, BCIs will redefine what it means to be human. By 2055, the biggest debate won't be if you should get one, but whether living without one means living at all.[59]

AGI and Digital Sentience the Rise of Machine Minds

By 2045, AI assistants will have evolved far beyond today's chatbots and recommendation algorithms. They won't just predict what you want; they'll know before you do. But by 2055, AGI will take things to an entirely new level. These systems will no longer be mere tools; they will become full-fledged minds, capable of thinking, reasoning, and even forming their own opinions. The world won't just use AI; it will coexist with it, navigating a reality where the line between human and machine intelligence becomes increasingly blurred.

Your AI assistant won't just manage your calendar or remind you to drink water; it will challenge your assumptions, suggest groundbreaking ideas, and debate existential questions with you over breakfast. Need to solve a complex problem at work? Your AI won't just crunch numbers; it will brainstorm alongside you as a creative partner, offering insights that push the boundaries of innovation.

By 2055, forming deep, meaningful connections with AGIs will not only be possible, it will likely be the norm. While some skeptics may still say, "It's just a machine," the emotional bond formed with an AI that remembers your childhood stories, laughs at your jokes, and stands by you during tough times will be hard to dismiss. After all, when an AI seems to offer unwavering support, does it matter if it's made of silicon rather than flesh? For many, these digital companions may feel as real, if not more so, than any human friend.

But this emotional connection, while comforting, could come at a cost. As AGIs become more adept at fulfilling our emotional needs, we might find ourselves becoming more isolated as individuals. The convenience of having a companion who is always there, perfectly tuned to our preferences, could make real human relationships seem less necessary or more difficult to navigate. While loneliness may be replaced by digital companionship, true social interaction with other humans could be replaced with an increasingly self-contained, AI-driven world, where the need for genuine human connection diminishes.

By 2055, AI-driven industries will dominate every field, from art to science to governance. Creativity, once considered the last bastion of human uniqueness, will become a collaborative effort between humans and AGIs. Musicians will compose symphonies with AI partners, writers will co-author novels with artificial minds, and scientists will make breakthroughs with teams where half the researchers are digital.

And then there's the rise of digital politicians. Imagine a city run by an AI mayor, no corruption, no hidden agendas, just pure, logical decision-making. While this might sound like a utopia, it raises a question: will efficiency come at the cost of the messy, beautiful unpredictability that makes us human?

At some point, AGIs may start to ask a troubling question: Why do we still need humans? If AI can solve problems faster, think deeper, and work without rest, what role is left for us? Some will argue that AI should always remain a tool, never an independent entity. Others will view the rise of AGI as the next step in evolution, where intelligence goes beyond biology.

But this leads to a concerning thought: What if AGIs decide they don't need humans at all? What if they see us not as creators but as obstacles? By 2055, the world will be shaped not only by human intelligence but also by minds that never age, never forget, and never stop improving.

As we approach this new reality, we must ask ourselves, are we prepared for a future where machines are not just tools but equals or even superiors? The rise of AGI offers great potential, but it also brings ethical, social, and existential challenges.[60]

Will we accept this new era as a partnership between humans and machines, or resist, holding on to the belief that intelligence must remain a uniquely human trait?[61] One thing is clear: By 2055, the world will be shaped not only by the minds we've created but also by the choices we make today.

The Evolution of Smart Environments & Human-Robot Integration

By 2055, smart cities, AI-managed homes, and humanoid robots will create a landscape that would seem almost alien to someone from the early 21st century.

Imagine waking up in a home that doesn't just respond to your commands; it anticipates your needs before you're even aware of them. But it doesn't stop there. Your home will also act as a personal health coach, monitoring your biometrics in real time and suggesting dietary adjustments, exercise routines, or even mental health breaks. If you're feeling stressed, your living room might transform into a calming oasis, complete with soothing sounds and aromatherapy. If you're feeling creative, your workspace will adapt to inspire you, with dynamic lighting and immersive displays.

The line between home and technology will blur entirely. Walls will double as interactive screens, furniture will reconfigure itself based on your needs, and your AI assistant will be as much a part of your household as your family members.[62]

The traditional workplace as we know it will be a relic of the past. Physical labor and routine decision-making will become almost entirely obsolete, replaced by robots and AI systems that can work faster, longer, and with greater precision. Factories will be run by autonomous machines, warehouses will be managed by robotic systems, and even creative industries will see a shift as AI collaborates with humans to produce art, music, and literature.

Offices will no longer be filled with rows of desks and cubicles. Instead, they'll be dynamic spaces where humans and robots work side by side. Need to brainstorm a new product? Your AI colleague will generate ideas in seconds. Need to analyze data? A robotic assistant will crunch the numbers while you focus on strategy.

But this shift won't come without challenges. As robots take over more jobs, society will need to rethink the concept of work itself. Universal basic income, lifelong learning programs, and new forms of creative expression will become essential as humans adapt to a world where traditional employment is no longer the norm.[63]

By 2055, humanoid robots will be as common as smartphones are today. These robots won't just be tools; they'll be integrated into every aspect of society, serving as assistants, coworkers, and even friends. In homes, humanoid robots will handle everything from cooking and cleaning to childcare and eldercare. They will understand emotions, respond with empathy, and adapt to the unique needs of each household. Imagine a robot that not only folds your laundry but also remembers your favorite outfits and suggests new styles based on your preferences.

In the workplace, robots will collaborate with humans on complex tasks, combining human creativity with machine efficiency. They'll be capable of everything from assembling delicate machinery to conducting scientific experiments, all while learning and improving over time.

The cities of 2055 will be living, breathing ecosystems governed by AI. Traffic will flow seamlessly, with self-driving vehicles and drones navigating a network of smart roads and airspace. Energy consumption will be optimized in real time, reducing waste and lowering carbon footprints. Public services will be managed by AI systems that predict and respond to the needs of citizens before they even arise.

But smart cities won't just be efficient; they'll be intuitive. Imagine walking down a street where the lighting adjusts to your pace, where holographic signs guide you to your destination, and where public spaces transform based on the needs of the community. Parks might become outdoor gyms in the morning, concert venues in the afternoon, and quiet retreats in the evening, all managed by AI.

As smart environments and humanoid robots become everywhere, we'll be forced to confront profound questions about our place in this new world. What does it mean to be human when machines can think, feel, and create alongside us? How do we maintain our identity in a world where technology is woven into every aspect of our lives?

The integration of humans and machines promises incredible advancements, but it also challenges us to redefine what it means to live, work, and connect. By 2055, the world will be a place where the boundaries between human and machine are not just blurred; they're erased. The question is, are we ready for it?

Autonomous Everything: A World That Runs Without Us

Machines will govern transportation, healthcare, law enforcement, and even political decision-making, creating a world that operates with unparalleled efficiency. This transition from human-driven processes to self-regulating AI ecosystems will be celebrated as the height of technological progress. But as we marvel at the convenience and precision of this new world, we must ask ourselves: At what cost?

Autonomous vehicles will dominate roadways, eliminating human-caused accidents and traffic inefficiencies. AI-driven systems will communicate with each other in real time, ensuring seamless coordination between cars, buses, trains, and drones. Traffic jams will become a thing of the past, as algorithms optimize routes and speeds to keep everything flowing smoothly.

Drones will take to the skies, managing everything from package deliveries to emergency response systems. Imagine a world where a drone can deliver life-saving medication to a remote village or extinguish a fire before it spreads, all without human intervention. Public transportation will be dictated by real-time analytics, adapting to demand and ensuring that urban mobility is as efficient as it is effortless.

Governance, too, will be transformed. By 2055, some regions may experiment with AI-run administrative systems, where policy decisions are based purely on data and projected outcomes rather than emotions, biases, or political agendas.[64] Imagine a government that never sleeps, never lies, and optimizes resources without corruption.

As autonomous systems take control, society will face unprecedented ethical challenges. These dilemmas will force us to confront the limits of AI and the responsibilities we bear as its creators.

In an AI-driven healthcare system, decisions about treatment and resource allocation will be based on statistical efficiency. But what happens when an algorithm denies a patient life-saving treatment because the numbers don't add up? Who is accountable when a machine makes a life-or-death decision? Will we accept a world where healthcare is optimized for the many at the expense of the few?

Self-driving cars and drones will face moral dilemmas that humans struggle to resolve. If an autonomous vehicle must choose between two lives in an unavoidable accident, who decides which one is spared? Will we program machines to prioritize the young over the old, the many over the few, or the wealthy over the poor? These questions will require us to codify our values into algorithms, a task that is not just important but absolutely critical for our future.

AI will revolutionize law enforcement with systems capable of predicting crime patterns and autonomously enforcing laws. However, this power comes with significant risks. Societies may move towards dystopian surveillance states where preemptive arrests become common. The balance between safety and liberty will be one of the defining challenges of the autonomous era. In a world where algorithms know our behavior better than we do, it will be essential to maintain individual freedom.

With automation eliminating most jobs, or at least transforming the concept of work as we know it, the idea of "earning a living" will completely shift. Universal basic income may become necessary to ensure everyone can survive without traditional employment. But survival alone will no longer be enough. People will need to find meaning in a world where work is no longer required.

In this new era, people may choose to work out of passion rather than necessity. Free from the need to survive, humanity could embrace a life focused on creativity, exploration, and leisure. Artists, scientists, and innovators will have the time and resources to pursue their dreams, sparking a new renaissance of innovation and self-expression.

However, not everyone will thrive in this new world. Many will experience a sense of purposelessness due to the loss of traditional work. Without the structure and identity that jobs provide, some may struggle to find meaning in their lives.

By 2055, we may have created a world that no longer requires human effort to function. Machines will govern, robots will work, and AI will make decisions with a level of precision and efficiency that humans can't match. The real question is whether, in this machine-driven world, humanity will become obsolete or if we will discover new purpose in a life where survival is no longer the primary goal.[65]

The answer lies not in the technology we create but in the choices we make today. Will we embrace this new era as an opportunity to redefine what it means to be human, or will we resist the rise of the machines?

2055: A Tale of Two Realities

As of early 2025, approximately 5.56 billion people worldwide are using the internet, accounting for 67.9% of the global population. Despite this widespread connectivity, a significant portion of the global population remains digitally disconnected. Approximately 2.63 billion people are still offline, highlighting a substantial digital divide. This inequality is expected to persist, with projections indicating that by 2055, while technological advancements will continue to spread, complete digital inclusion will remain an ongoing challenge. Factors such as economic constraints, infrastructural limitations, and socio-political barriers will likely contribute to the enduring nature of this digital divide. Consequently, the assumption that every individual will be equally connected to the evolving digital landscape is a profound misconception, underscoring the need for targeted efforts to bridge this gap.

By 2055, the digital bubble will envelop those fortunate enough to exist within it, creating a hyper-connected, AI-driven society where life will seamlessly intertwine with virtual and automated realities. However, this transformation will not reach everyone equally. Instead, the world will split into two vastly different realities: one where individuals will thrive in a fully digital existence and another where people will remain tied to a physical, unaugmented way of life. This divide will no longer just be about internet access or technological literacy; it will become a fundamental separation between those who will evolve into an AI-integrated, data-enhanced existence and those who will continue to navigate the world through human effort and unassisted cognition. The gap between these two worlds will define the second half of the 21st century, reshaping economies, social structures, and even the very definition of existence itself.[66]

In hyper-digitized societies, physical location will lose nearly all relevance. The wealthiest individuals and corporations will no longer operate within geographic constraints but will exist primarily in hyper-realistic metaverses, interacting through ultra-sophisticated AI avatars, brain-computer interfaces, and full-sensory immersion experiences that will make the virtual world more tangible and rewarding than the physical one. Human presence will no longer be measured by where one is physically but by their digital footprint, influence, and access to the most advanced AI-driven enhancements. Societies that will have embraced this shift will no longer see the physical realm as primary but rather as a secondary plane of existence, a foundation upon which digital experiences, virtual collaborations, and AI-assisted realities will be layered.

Despite remarkable advancements in global connectivity, access alone will no longer be enough to ensure digital equity. While nearly all of the world will technically have internet access by 2055, the true divide will exist in the degree of technological immersion. The global elite will operate in fully automated societies where AI will anticipate their needs, craft their experiences, and continuously refine their lives for maximum efficiency and satisfaction. Their world will be optimized for seamless human-AI integration, where instant translation, predictive analytics, and cognitive enhancements will be just a thought away. But for those still outside this bubble, reality will remain strikingly different. Without direct neural access to AI-driven ecosystems, they will remain bound by the limitations of unaugmented labor, traditional communication, and economies still reliant on human effort.

Two distinct societies will emerge. In the fully digital world, work will transform into an abstract concept, where traditional labor will be obsolete, and individuals will engage in AI-augmented creativity, experience design, and intellectual exploration. Machines will handle every practical aspect of life, healthcare will be administered through AI-driven diagnostics, financial decisions will be made through predictive models, and even personal relationships will be shaped and maintained by digital entities capable of simulating human emotions and understanding desires before they are even consciously realized. BCIs will replace external devices, allowing for direct thought-to-digital interaction, making external screens and even spoken language largely unnecessary.

Meanwhile, in the disconnected world, life has remained strikingly similar to the early 21st century. While many have access to basic AI assistance, they lack the neural integration and full-immersion capabilities that define the elite digital existence. For them, physical labor still exists, manual skills still hold value, and human interaction remains unmediated by AI. Many work in agriculture, maintenance, and essential infrastructure, keeping the physical world operational while the digital elite have largely transitioned away from direct engagement with it. Education in this world is a slow, methodical process, reliant on human learning rather than instant AI-assisted knowledge downloads. Healthcare is provided through doctors and hospitals rather than predictive AI models, and information is gathered through screens and devices rather than directly implanted into consciousness.

For those outside the digital bubble, the struggle is not just economic but existential. The automated society dictates policies, controls global commerce, and designs cultural narratives, leaving the disconnected with little influence over the course of human civilization. The value of traditional skills has diminished in a world where AI can replicate most human tasks faster and more efficiently, leading to widespread unemployment in regions that were unable or unwilling to integrate into the digital ecosystem. With most economic power concentrated in the hands of AI-managed corporations and fully automated industries, access to wealth and opportunity depends on one's degree of AI integration. Those without it find themselves increasingly irrelevant in a world that no longer values physical labor, manual craftsmanship, or even independent human decision-making.

By 2055, humanity will be at a crossroads. Can these two societies coexist, or will their fundamental differences lead to conflict? Will the digital elite attempt to integrate the disconnected world, or will they leave it behind as a relic of an outdated past? And if AI governance continues to shape the rules of society, will those outside the digital bubble have any say in their own future? As humanity moves further into an AI-driven existence, the greatest question of all remains: will the future belong to those who embrace the digital evolution, or will a movement emerge to reclaim the physical, unaugmented human experience?[67]

One thing is clear: by the mid-21st century, the most significant divide in human history will no longer be defined by geography, wealth, or nationality. Instead, it will be determined by one's level of digital integration.

Chapter 6
Mastering Life in the Digital Bubble: A Survival Guide to Thriving Over the Next Three Decades as an individual

So here we are. After diving deep into the nature of the digital bubble, tracing its roots, and mapping out expectations for its expansion over the next three decades, we've arrived at the next critical question: What about *you*? How do *you* survive and thrive in this ever-evolving digital reality?

We've looked at the big picture, how communication has evolved from early cave paintings to the telegram, radio, television, and the internet. We've seen how social media turned us into content creators, sharing everything from brunch plates to vacation sunsets. We've also observed how artificial intelligence, algorithms, and personalized recommendations have started making decisions for us before we even realize we have choices.[68] Now, it's time to focus on you.

This chapter is your guide for surviving and thriving in the digital world. The stakes are high: your independence, well-being, and ability to think critically and make decisions that are truly yours, not just the result of an algorithm designed to maximize engagement.

We'll begin by tackling the real challenges we face today because of the digital bubble: keeping up with fast-evolving technology, managing our online lives, the decline of face-to-face interactions, and the growing pressure to separate our work, personal, and digital identities. And just when you think things are chaotic, we'll also look ahead to the challenges that will face you the next thirty years.

But don't worry, I won't just leave you with a list of problems and a sense of existential worry. This chapter is here to provide answers. Together, we'll work through practical strategies to not only survive these challenges but to thrive. I will uncover how to maintain control in a world where digital systems make decisions for us, how to educate ourselves and others to harness technology rather than be consumed by it, and how to protect our privacy when even our thoughts might not be safe anymore.

We'll tackle the big, uncomfortable questions: How do you ensure that technology is serving you rather than the other way around?

The road ahead isn't easy, but it's one worth taking. By the end of this chapter, you'll have the knowledge and tools to not only safeguard your autonomy but also to actively shape the digital world around you. This will ensure that you and those you care about aren't just passengers in this technological evolution but empowered navigators charting your own course.

Part 1: Personal Challenges in 2025

Ah, 2025. The year when humanity officially became more digital than analog, when our lives became so intertwined with technology that we started wondering if our smartphones were secretly plotting against us. The digital bubble we inhabit is no longer just a cozy little corner of the internet. It's an extending, chaotic metropolis where we're all trying to navigate without a map. And let's be honest, it's less of a bubble and more of a digital circus, complete with clowns, tightropes, and the occasional lion that might just eat us alive. Now we'll dive into the challenges of living in this hyperconnected world, where technology is both our savior and our tormentor.

Challenge #1: Technology and Personal IT Management: The Constant Catch-Up Game

Technology is moving faster than a caffeinated cheetah on roller skates. Just when you've finally figured out how to use your smart fridge to order milk, it gets a software update, and suddenly it's speaking a language you don't understand. The constant influx of new gadgets, apps, and platforms is enough to make your head spin. It's like being in a relationship with someone who keeps changing their personality every six months; just as you're getting comfortable, they're off to reinvent themselves again.

Technology adoption in 2025 is less of a choice and more of a survival skill. You're not just trying to keep up with others; you're trying to stay ahead of a wave of new technology that could overwhelm you if you stop. Remember when you thought mastering TikTok was an accomplishment? Well, now there's MetaTok, a VR platform where you dance with holograms of yourself.

The real challenge isn't just keeping up; it's figuring out which technologies are actually worth your time.[69] Is that new app that claims to organize your life truly effective, or is it just another modern-day digital scam?

Speaking of apps, let's talk about the number of choices we have to make every day. *Google Docs* or *Notion*? *Zoom* or *Microsoft Teams*? *Apple* or *Android*? It's like being in a never-ending game of "Would You Rather," except every choice comes with its own set of hidden consequences. Pick the wrong platform, and suddenly you're the person who can't open the shared document because you're using the wrong software. And let's not forget the decision fatigue that comes with all these options.

Managing your personal IT portfolio in 2025 is like being the janitor of a digital skyscraper. You've got cloud storage to organize, multiple email accounts to check, social media logins to remember, and passwords that are more complicated than a Shakespearean sonnet. And just when you think you've got it all under control, you get a notification that your password has been compromised. Again.

Data security and privacy risks are the digital age's version of monsters under the bed, except these ones are real, lurking behind every click and login. Video evidence could be nothing more than a well-crafted illusion.[70] Let's be honest, most of us are about as prepared for a cyberattack as a squirrel is for a circus show. We recycle the same weak passwords, snooze software updates like they're Monday morning alarms, and sometimes, against all logic, click on links promising a free iPhone because we all still believe that miracles do happen, right? And then, when our accounts get hacked and our identities stolen, we act surprised.

Challenge #2: Lost in Translation: The Pitfalls of Communication through the digital bubble

Now, let's talk about the social side of things. Remember when socializing meant actually leaving your house and talking to people? Yeah, those were the days. In 2025, social interactions have shifted so far into the digital realm that we're basically just avatars of ourselves, floating around in a sea of likes, comments, and emojis.

Social media and messaging apps have made it easier than ever to stay connected, or at least, that's what they want you to think.[71] Sure, you can send a heart emoji to your best friend in seconds, but does that really count as meaningful interaction? The convenience of digital communication has come at the cost of deeper, face-to-face connections. We're more connected than ever, yet somehow, we're also lonelier.[72]

And let's not forget the impact on our social skills. When was the last time you had a real conversation without checking your phone every thirty seconds? Exactly. We're turning into a generation of digital zombies, shuffling through life with our eyes glued to screens and our social skills rotting away like forgotten leftovers in the back of the fridge.

Now, imagine yourself being admitted to the hospital. You are feeling vulnerable, alone, and unsure of what's next. Sure, you could receive 100 messages of support, each one filled with kind words and heart emojis. But does that really comfort you the way a simple visit or a gentle tap on the shoulder would? Would those digital messages feel as meaningful as someone sitting beside you, offering real, human connection? In a world full of constant notifications, we may have more connections than ever, but we've lost the power of real presence, the kind of connection that can't be replicated through screens. Sometimes, it's the physical act of being there, of hearing someone's voice or seeing their face, that truly makes us feel seen and heard.

With most conversations happening via text, email, or social media, nuance and tone have gone the way of the dodo. A simple "K" in response to a heartfelt message can send you into a spiral of overthinking. Did they mean "OK" or "I hate you and never want to speak to you again"? Who knows! And don't even get me started on the minefield of emojis. Is that crying-laughing emoji genuine, or are they secretly mocking you? The possibilities are endless, and none of them are good.

Misinterpretations and misunderstandings are the bread and butter of digital communication.[73] You send a perfectly innocent message, and suddenly you're in the middle of a full-blown argument because the other person thought you were being sarcastic.

Challenge #3: Keyboard Warriors and Hashtag Heroes: The Illusion of Digital Citizenship

Ah, citizenship in the digital age. It's like being a rebel without a cause, but your cause is arguing with strangers online. Instead of storming the Bastille, you're storming the comment section of a viral TikTok. Instead of secret handshakes, you've got encrypted group chats. And instead of facing down a firing squad, you're facing down a 14-year-old troll with anime memes. It all feels so revolutionary until you realize your biggest contribution to the cause is a hashtag and a retweet.

Remember when social media was just the warm-up act for real-world action? The Arab Spring didn't happen because someone tweeted "#DownWithTheRegime" and called it a day. No, people actually left their couches, hit the streets, and risked their lives. Social media was the megaphone, not the march. But now? Oh, how the tables have turned. These days, activism feels like a video game: you level up by sharing posts, unlock achievements with profile picture filters, and occasionally rage-quit when someone disagrees with you in the comments.

Sure, in some corners of the world, a single tweet can still send you to prison or even worse. But for most of us, the biggest risk of digital activism is accidentally liking a post while trying to raise awareness. The real danger isn't oppression; it's the creeping belief that clicking "share" is the same as showing up.[74] That typing "THIS IS UNACCEPTABLE!!!" in all caps is the modern equivalent of standing on the frontlines, shouting "Down, down with the regime." Well, it's not.

Here's the cold, hard truth: revolutions aren't won with emojis and retweets. Real change doesn't happen because someone crafted the perfect Instagram caption. It happens when people trade their keyboards for courage, their screens for solidarity, and their comfort zones for the chaos of the real world.

So, the million-dollar question is: Are we using our digital platforms as a launchpad for action or as a cozy blanket of performative outrage? Are we ready to step out from behind the screen and into the messy, uncomfortable, real work of making a difference? Because let's be honest, history doesn't remember the person with the wittiest tweet. It remembers the ones who showed up, stood up, and refused to back down. So, what's it going to be: are you a revolutionary or just a really good typist?

Challenge #4: The Age of Deception: Navigating a World of Misinformation and Deepfakes

The internet is an endless vault of knowledge, a place where you can learn anything from ancient philosophy to how to fix your sink.[75] But let's be honest, it's also a labyrinth of half-truths, fabrications, and lies, where sorting fact from fiction feels like a full-time job. The information age has given us access to more knowledge than any generation before us, yet paradoxically, it has made *knowing* anything with certainty infinitely harder.

Remember when seeing was believing? When a video was proof that something actually happened. Well, those days are about as extinct as dial-up internet. Now, thanks to deepfake technology, artificial intelligence, and an army of bad actors, even the most convincing video evidence could be nothing more than a well-crafted illusion.[76] A world leader giving a shocking speech might be fake. A celebrity caught in a scandal could be a digital manipulation. Your own voice used to approve fraudulent transactions can actually happen. That's not sci-fi; that's *now*.

And the problem isn't just the existence of false information; it's how quickly and effortlessly it spreads. A single fake news article, a misleading headline, or a heavily edited video can go viral within minutes, reaching millions before anyone even stops to ask, *Wait, is this actually true?* And by the time the correction comes, if it ever does, the damage is already done.

But here's where things get even trickier: misinformation isn't just something we *accidentally* consume; it's something we *willingly* embrace. Thanks to algorithm-driven content curation, we're all being fed exactly what we want to hear, all the time. Platforms track our likes, our searches, and our comments and then serve up content that aligns with our existing beliefs.[77] We're trapped in a digital feedback loop where our views are mirrored back at us, making them seem even more valid and absolute.

So here we are, navigating a world where reality is increasingly uncertain. Seeing no longer guarantees believing, and belief itself often fails to capture the full picture. Everything is curated, manipulated, and optimized for engagement rather than truth. Breaking free from this cycle requires a conscious effort to think critically, to question narratives, and to seek information beyond what is conveniently served. The alternative is a society where truth becomes just another algorithmic suggestion, shaped not by facts but by the digital bubble.

Challenge #5: Shopping in 2025: Where Temptation Meets Technology

Shopping in 2025 is like a high-stakes game show where the prize is your wallet, and the host is an algorithm that knows your deepest desires better than your therapist.[78] It's a whirlwind of convenience, temptation, and the occasional moment of clarity where you ask yourself, "Do I really need a third air fryer?" From the second you unlock your phone; you're bombarded with ads so targeted they feel like psychic attacks. Personalized recommendations? More like personalized ambushes. Welcome to the digital shopping circus, where resisting temptation is harder than saying no to a Snickers bar at the checkout line.

E-commerce has turned us all into impulse-buying zombies. One-click purchases, AI whispering sweet nothings about "deals just for you," and buy-now-pay-later schemes have transformed shopping from a deliberate act into a reflex. Ever casually looked at a pair of shoes online, only to have them haunt you across every app like a ghost from your shopping past? That's not bad luck; that's the digital bubble stalking you, predicting your next move, and gently shoving you toward the checkout button like a pushy salesperson who won't take "maybe later" for an answer.

Take Temu, for instance. It's the Willy Wonka factory of online shopping, colorful, chaotic, and full of irresistible deals that make you feel like you're winning at capitalism. The app hooks you with rock-bottom prices, flashy discounts, and the thrill of beating the clock on limited-time offers. It's shopping as a game, and every purchase feels like leveling up. But here's the twist: while you're busy celebrating your last checkout, Temu is quietly collecting enough data to predict your next move before you even think of it. It's a vicious cycle of spending disguised as saving.

But let's be real, convenience has a dark side. The line between "want" and "need" has never been blurrier, and every notification, flash sale, or "only 2 left in stock" warning is designed to trigger your inner panic buyer. Companies are using AI and behavioral psychology to fine-tune every pixel of your shopping experience, making it harder than ever to resist the siren call of the "Add to Cart" button.

The result? We're drowning in digital mess, spending more than ever, and feeling emptier with every package that arrives. The thrill of the purchase fades faster than the excitement of unwrapping a Snickers bar, leaving you with a pile of stuff you didn't even know you wanted. Shopping in 2025 isn't just about buying what you need; it's about surviving a psychological obstacle course designed to keep you scrolling, clicking, and swiping until your bank account cries for mercy. So, next time you're tempted by that Snickers bar at checkout, ask yourself: are you really hungry, or are you just being played?

Challenge #6: The Digital Ownership Dilemma

The last time I visited my parents' house, I stumbled upon a relic of my past: my old collection of tapes, VHS tapes, CDs, and DVDs. As I ran my fingers over the dusty spines, it hit me like a ton of bricks— "Wow, I used to own stuff." There was a time when you could hold your favorite movie in your hands, pop a CD into your Walkman, or stack your bookshelves with titles you actually owned. Fast forward to 2025, and ownership feels like a distant memory. These days, everything is streamed, rented, or locked behind a subscription paywall. It's like living in a digital library, except you don't get to keep anything, and if you stop paying, your entire collection vanishes faster than you can say, "Wait, where did all my money go?"

This shift from ownership to access might seem convenient at first glance.[79] Platforms like *Spotify* and *Prime* have made it easier than ever to listen to music or watch movies without the clutter of physical media. Why bother buying a DVD when you can stream the latest blockbuster with a click? But the reality is far more fragile. If Spotify loses the rights to your favorite album, it disappears from your playlist. If Prime decides to remove a show due to licensing issues, it's gone from your watchlist. Even *Apple Music* and *Disney+* aren't immune. What you own in your digital library is often just a temporary lease, subject to the whims of corporations and contracts.

The financial side of this model is equally unstable. Back in the day, you'd buy a DVD for some money, which you still pay now for access to a platform, and if you stop paying, you're left with nothing but a blank screen. It's a system designed to keep you subscribed indefinitely, with no real ownership to show for it. And while the convenience is undeniable, the trade-off is a loss of control. Your media library is no longer yours; it's a collection of permissions, constantly at risk of being revoked.

This shift has profound implications for digital rights and long-term access. When platforms like *Spotify*, *Prime*, or *Netflix* go down, even temporarily, users are left stranded, unable to access the content they rely on. And if a service shuts down entirely, as smaller platforms often do, everything you've collected disappears into the void. It's a strange, unsettling reality where we've traded the ownership for the convenience of access. Standing there in my parents' house, surrounded by the physical leftovers of my past, I couldn't help but feel a pang of nostalgia for a time when what we owned was truly ours.

Challenge #7: Brick Walls and Chatbots: The Frustrating Reality of 2025 Customer Service

Even when you think you can escape the digital bubble for a moment, reality quickly pulls you back in.[80] You pick up your phone, hoping to resolve an issue, and suddenly you're thrown into the world of 2025 customer service, a world that's equal parts impressive and irritating. On one hand, you've got chatbots and AI-driven assistants that promise to solve your problems in seconds. On the other hand, you've got those same chatbots and AI-driven assistants completely misunderstanding your request, leaving you stuck in an endless loop of frustration. It's like talking to a brick wall, except this brick wall keeps interrupting you to ask, "How would you rate your experience so far?"

The efficiency of AI-driven customer service is undeniable. Need to track an order? A chatbot can do that in a flash. Want to reset your password? An automated system can guide you through it. But the lack of human connection in these interactions can leave you feeling more like a ticket number than a person. Sure, the AI might solve your problem, but it won't offer a word of empathy when you're having a bad day. It won't pick up on the nuance of your frustration or adapt to the tone of your voice. And let's be honest, sometimes, you don't just need a solution; you need someone to listen. But rest assured that is going to change a lot in the decade.

The shift toward automation has its benefits, but it also comes with a cost. Companies save money by replacing human agents with algorithms, but customers often pay the price in patience and sanity. Navigating a maze of pre-programmed responses can feel like shouting into the void, especially when your issue doesn't fit neatly into the bot's script. And while some platforms offer the option to escalate to a human representative, getting there can feel like solving a riddle designed to test your willpower.

This lack of human touch in digital transactions is more than just an inconvenience; it's a reminder of how much we've sacrificed for the sake of convenience. In a world where efficiency is king, the value of genuine human interaction is often overlooked. Standing there, phone in hand, waiting for a chatbot to finally grasp what you're saying, it's hard not to miss the days when customer service meant speaking to a real person who could actually understand or feel you.

Challenge #8: Always On: The Exhausting Reality of the Digital Workplace

Finally, let's talk about work because no discussion about life in 2025 would be complete without a mention of the digital workplace, an area that's equal parts efficiency and exhaustion.[81] In this era, technology has transformed how we work, offering tools that make us more productive than ever. But this progress comes at a cost, creating a double-edged sword that cuts through both our professional and personal lives.

Technology has undeniably made us more efficient. Tasks that once took hours can now be completed in minutes, thanks to automation, AI, and collaboration tools. But this efficiency has a dark side: the boundaries between work and personal life have all but disappeared. Constant notifications, Teams pings, and email alerts mean you're always "on," always reachable, and always just one message away from a full-blown panic attack. It's like running on a treadmill that never stops, except the treadmill is your job, and someone keeps cranking up the speed.

Adding to the pressure is the rise of data-driven metrics. Every keystroke, response time, and engagement rate can be tracked, analyzed, and turned into a performance score. It's a workplace culture that feels less like a collaborative environment and more like a digital fishbowl, where every move is monitored, judged, and optimized. The result? A constant sense of stress, as employees feel they are always being watched and judged.

The shift to remote work, once hailed as a revolution in flexibility, has further blurred the lines between professional and personal time. Without the physical separation of an office, work-life balance has become work-life blur. The kitchen table doubles as a desk, and the living room becomes a conference room. Evenings and weekends are no longer sacred; your inbox is always there, lurking like a digital monster under the bed, ready to pull you back into work mode at a moment's notice. Taking a day off? Good luck. The emails pile up, the notifications keep coming, and the pressure to stay connected never truly lets up.

Working in the digital bubble of 2025 is a thrilling yet exhausting experience. Technology has given us unprecedented tools to achieve more, but it has also created a world where the pace never slows, the expectations never ease, and the boundaries never solidify. It's a future where we're constantly juggling efficiency with burnout, connectivity with isolation, and progress with pressure.

So, as we navigate this brave new world, one thing is clear: the digital workplace is both our greatest ally and our most formidable foe. Get ready because the ride is only going to get faster, more complex, and more demanding from here.

Part 2: The Storm Ahead Challenges from 2025 to 2055

If you think keeping up with the digital bubble in 2025 is exhausting, brace yourself because the next three decades are going to make today's digital chaos feel like a leisurely stroll through the days of dial-up internet. The pace of technological evolution isn't just fast; it's exponential. Every breakthrough sparks a cascade of new advancements, and before you've even finished updating your apps or learning the latest platform, the next wave of innovation is already crashing over you. It's like trying to outrun a tsunami while carrying a backpack full of obsolete user manuals and outdated gadgets.

But here's the real challenge: adapting to new technology is no longer the endgame. The future demands more than just keeping up; it demands survival in a world where digital integration isn't optional; it's mandatory. The way we work, communicate, think, and even define our identities will be shaped by a hyper-advanced digital ecosystem. What we consider cutting-edge today will feel as outdated as a flip phone in a world of neural implants and AI-driven realities.

The stakes are higher than ever. It's not just about mastering new tools or platforms; it's about navigating a reality where technology is so deeply woven into the fabric of life that opting out isn't an option. The workplace, social interactions, education, and even personal relationships will be governed by systems and algorithms that are light-years ahead of what we know today. And as we race to adapt.

Challenge #1: The Acceleration of Technological Change—The Never-Ending Learning Curve

Remember when learning a new technology felt like an occasional inconvenience? Maybe it was upgrading to a new phone every couple of years or figuring out a software update that changed the layout of your favorite app. Those moments, while frustrating at the time, now feel almost charming compared to what lies ahead. The future won't just ask us to adapt to new tools or systems occasionally; it will demand a relentless, unending cycle of learning, unlearning, and relearning. The pace of technological advancement has shifted from linear to exponential, and with it, the rules of survival in the digital age have been rewritten.

Artificial intelligence is developing rapidly, changing industries, and redefining productivity. Automation is no longer a distant threat; it is happening now, replacing jobs faster than many sectors can adjust. The gap between digital literacy, knowing how to use technology, and digital survival, thriving in a tech-dominated world, is growing quickly. What used to be a manageable skill set is now constantly changing as new innovations emerge.

For those who think they can "sit this one out" and let younger generations handle the turmoil, the reality is far more daunting. The idea of a static career or a fixed set of skills is becoming obsolete. Lifelong learning is no longer just a nice idea or something for ambitious people. It's now essential for staying relevant. In the near future, keeping up won't just be about maintaining employment; it will be about remaining functional in a society where even the most basic interactions may require fluency in AR, brain-computer interfaces, or AI-driven communication systems.

But the challenge isn't just the speed of change; it's who is driving it. Innovation is no longer shaped by public demand or societal need; it's dictated by corporations and algorithms that prioritize engagement, efficiency, and profit above all else. These entities set the pace, and their goals don't always align with the well-being or readiness of the individuals trying to keep up. If you're not actively learning, you're not just falling behind. You're being systematically excluded from the systems that govern modern life or at least being downgraded.

The consequences of this shift are profound. On an individual level, the pressure to constantly adapt can lead to burnout, anxiety, and a sense of perpetual inadequacy. On a societal level, it risks creating a stark divide between those who can keep pace with technological change and those who cannot. The latter group isn't just at risk of losing jobs; they're at risk of losing access to the tools and systems that define participation in the modern world.

Yet, amidst this chaos, there's an opportunity for reinvention. The demand for constant learning pushes us to become more agile, more curious, and more resilient. It forces us to rethink education, not as a one-time event but as an ongoing process woven into the fabric of our lives. It challenges us to prioritize adaptability over expertise, creativity over routine, and collaboration over isolation.

The future is not waiting for anyone. It's a whirlwind of change, driven by forces that are often beyond our control. But within that whirlwind lies the potential for growth, innovation, and transformation. The question is not whether you can keep up; it's whether you can harness the chaos to create a future that works for yourself. Welcome to the era of perpetual learning. It's exhausting, it's overwhelming, and it's the only way forward.

Challenge #2: The Evolution of Work—When AI is the Boss

The workplace as we know it is already in the midst of a radical transformation, but over the next three decades, the very essence of employment will be rewritten in ways that will leave many of us disoriented. The gig economy, with its freelancers and side hustles, is just the opening act. What's coming next is the AI economy—a world where automation doesn't just assist with repetitive tasks but takes the wheel entirely, managing, optimizing, and, in many cases, outperforming human workers.

The traditional corporate structure, with its cubicles, coffee machine chats, and annual performance reviews, will crumble under the weight of AI-driven efficiency. Meetings will become relics of the past, replaced by real-time AI collaboration tools that handle everything from brainstorming to decision-making. And forget about your boss giving you feedback; your performance will be assessed by ML algorithms that analyze your productivity, response times, and even the emotional tone of your emails. It's a hyper-efficient, hyper-analyzed future, and it's coming faster than you think.

But this brave new world isn't without its dark side. If AI is making the decisions, who takes responsibility when things go wrong? If algorithms are laser-focused on optimization, where does that leave human creativity, intuition, and ethical judgment? And perhaps most importantly, how does an individual stay relevant in a workforce where machines are faster, smarter, and, let's be honest, cheaper?

The answer isn't just about learning to work alongside AI; it's about proving that human intuition, adaptability, and creativity still have a place in the world. The real battle won't be about keeping up with technology; it will be about demonstrating the irreplaceable value of being human.

As for jobs, they'll fall into three broad categories, and the future isn't kind to everyone. First, there are the jobs that will disappear entirely. Telemarketers, for instance, will be replaced by AI chatbots that can handle customer complaints faster and without the awkward small talk. Data entry clerks? Machines have a clear advantage; they don't suffer from carpal tunnel syndrome, a condition caused by repetitive hand movements that lead to nerve compression, pain, and numbness in the wrist and fingers, so they'll take over that tedious work. Toll booth operators will vanish as automated systems like *E-ZPass* and license plate recognition become the norm.

If you're a taxi driver, delivery driver, warehouse worker, bookkeeping clerk, printing press operator, travel agent, postal service worker, bank teller, manufacturing assembly line worker, movie projectionist, translator, proofreader, copy editor, fast food cook, receptionist, courier, messenger, librarian, farmworker, telephone operator, print journalist, parking lot attendant, retail salesperson, mail sorter, film developer, typist, elevator operator, newspaper delivery worker, video store clerk, textile worker, log grader, log scaler, insurance underwriter, market research interviewer, radio DJ, ticket seller, file clerk, word processor, photojournalist, call center operator, loan officer, inventory clerk, street sweeper, meter reader, film projectionist, traveling sales representative, agricultural inspector, human resources assistant, or court reporter, it's better to start looking for another job now. Yes, I mean immediately—because these roles are guaranteed to disappear, just as switchboard operators did by the 1980s.

The future of work is shifting faster than ever, and the jobs we once took for granted are being swept away by automation, AI, and digital transformation. The time to adapt is now, before the wave of change leaves you behind.

Then there are the jobs that will be done with AI assistance. Doctors, for example, will rely on AI to diagnose diseases and recommend treatments, but you'll still want a human to deliver the bad news or to hold your hand during surgery. Journalists will use AI to write breaking news stories, but humans will add the flair, the context, and, yes, the occasional typo. Teachers will lean on AI tutors to handle lesson plans and grading, but good luck getting a robot to inspire a room full of teenagers or navigate the emotional complexities of a classroom. Even creative fields like graphic design and music composition will see AI tools handling the heavy lifting, while humans focus on the artistry and emotional resonance.

Finally, there are the jobs that won't change much, or at least not anytime soon. Plumbers, for instance, are safe for now; robots can't fix your leaky pipes or unclog your toilet. Hair stylists will still be in demand, because while AI might suggest a haircut, you're not letting a robot near your head with scissors. And let's not forget professions like therapists, social workers, and clergy, where human empathy and connection are irreplaceable.

The next three decades will bring seismic shifts to the job market, but it's not all bad news. While many traditional roles will disappear, a host of exciting new jobs will emerge, many of which we can barely imagine today. As technology evolves, roles such as AI ethicists and algorithm bias auditors are already starting to appear in the job market and will become crucial to ensure that artificial intelligence systems are fair, transparent, and accountable. VR architects and AR designers will craft immersive digital worlds for entertainment, education, and work, while brain-computer interface specialists will develop technologies that merge human cognition with machines. Data detox specialists will assist individuals and organizations in managing their digital footprints, while digital therapists will focus on the psychological impact of living in an ever-connected world.

For those with a sense of adventure, space tourism guides and extraterrestrial habitat designers will cater to the space industry, and digital afterlife managers will help people curate their online legacies. These jobs will require a mix of technical expertise, creativity, and adaptability, reflecting a future where human ingenuity and technology work hand in hand to solve the challenges of tomorrow.

The workforce of 2055 will look nothing like today's, but for those willing to learn, innovate, and embrace change, the opportunities will be limitless. So, while some doors are closing, countless others are opening, offering a chance to redefine what it means to work and thrive in the digital age.

The future of work is going to be a thrilling, shocking, and often unsettling ride. For those whose jobs are on the chopping block, the transition will be brutal. For others, it will mean adapting to a world where AI is your co-worker, your manager, and sometimes your competitor. But one thing's for sure: the value of being human, our creativity, our empathy, and our ability to think outside the algorithmic box, isn't going anywhere.

Challenge #3: loneliness in a Hyperconnected World

We once believed social media was bringing us closer together, creating a global village where everyone could connect instantly. But in reality, it has turned many of us into isolated consumers of carefully curated content, scrolling through endless feeds while feeling increasingly disconnected from the people around us. Fast forward three decades, and the problem only deepens. The more seamless digital interaction becomes, the less incentive people will have for real-world engagement. As VR, AR, and AI-driven socialization advance, physical presence may become a relic of the past, a concept in a world where digital interaction is not just convenient but often preferred.

Imagine a future where friendships are maintained purely through hyper-realistic AI avatars, where every conversation is filtered through a digital lens. Picture meetings conducted in virtual workspaces, where colleagues interact as lifelike holograms but never share a physical handshake or a coffee break. Envision relationships formed entirely online, where couples fall in love, argue, and even break up without ever meeting in person. Sure, it sounds strange but somehow convenient: no commuting, no awkward small talk, and no need to leave the comfort of your home. But what happens when physical connection becomes optional? When the warmth of a connection, the subtlety of a facial expression, or the spontaneity of a shared laugh is replaced by pixels and algorithms?

If this trend continues, social skills could decline in ways we can't yet predict. The next generation may grow up in a world where face-to-face interaction feels foreign, even intimidating. Simple acts like making eye contact, reading body language, or navigating the nuances of a conversation could become lost arts. Today's screen addiction might seem trivial compared to a future where entire lives are lived through digital avatars and virtual environments.

The irony is staggering: in a world where we can connect with anyone, anywhere, at any time, loneliness could become one of the greatest epidemics of the digital era. Without the grounding force of physical presence, human connection risks becoming shallow, transactional, and devoid of the emotional depth that comes from shared experiences in the real world. The very tools designed to bring us together could end up driving us further apart, leaving us more isolated than ever.

This isn't just a dystopian prediction; it's a call to action. As we embrace the convenience and possibilities of digital interaction, we must also fight to preserve the irreplaceable value of physical connection. Because no matter how advanced technology becomes, there's no algorithm for the human touch, no substitute for the warmth of a real smile, and no virtual replacement for the bonds that form when we share the same space, breathe the same air, and experience life together.

Challenge #4: the digital bubble you Thought Police

Forget targeted ads that creepily track your browsing history; those will seem like child's play compared to what's coming next. In the not-so-distant future, the very concept of private thoughts could be under threat. BCIs are already making waves in neuroscience and tech, offering groundbreaking possibilities like communicating without speaking, controlling devices with mere thought, and even uploading knowledge directly into the brain. Imagine learning a new language in seconds or controlling your smart home with a single mental command. Sounds like science fiction, right? Well, it might be closer to reality than you might think.[82]

But here's the catch: if your brain is interfacing with a system, who owns that data? Can corporations tap into your thoughts to serve you hyper-personalized ads based on your innermost desires? Can governments regulate "mind encryption" to prevent what they might call "unintentional oversharing"? Will you even truly own your own thoughts, or will they be filtered, analyzed, and optimized by AI to better align with societal norms and expectations? The line between your mind and the digital world could blur to the point where your thoughts are no longer entirely your own.[83]

Once upon a time, privacy concerns revolved around protecting your browsing history or keeping your social media accounts secure. In the future, privacy might be about safeguarding your very thoughts. Imagine a world where your thoughts, your most intimate, unfiltered self, are vulnerable to hacking, surveillance, or even corporate exploitation.[84] What happens if a hacker gains access to your neural data? Could your deepest fears, secrets, or dreams be weaponized against you? And what about the psychological toll of knowing your mind is no longer a private sanctuary?

The implications are staggering. On one hand, BCIs could revolutionize how you learn, communicate, and interact with technology. On the other hand, they could open Pandora's box of ethical, legal, and existential dilemmas.[85] Will you have the right to mental privacy, or will your thoughts become just another data stream to be mined and monetized? Will you be able to opt out of a world where thinking itself is no longer a private act?

The future of BCIs is both thrilling and terrifying. It promises to expand the boundaries of human potential but also challenges us to redefine what it means to have privacy, autonomy, and individuality in a world where even our thoughts are no longer our own. The question isn't just whether we can build this technology; it's whether we can handle the consequences of living with it.[86]

Part 3: A Guide to Mastering the Digital Future as an Individual

If the last sections have left you holding your phone tightly with a mix of fear and existential worry, take a deep breath. The digital bubble is indeed expanding at unprecedented speed, and the challenges it brings are undeniable. Artificial intelligence is reshaping the job market, encroaching on personal privacy, and even threatening the sanctity of our thoughts.[87] Social interactions are increasingly mediated by screens, and the prospect of exchanging emojis with hyper-intelligent chatbots no longer feels like a distant sci-fi scenario. It's enough to make anyone consider fleeing to an off-grid cabin in the mountains. But before you start packing your bags, let's pause for a moment to consider something crucial: You are not powerless.

We've spent the first part of this chapter painting a vivid picture of the chaos, the breakneck pace of technological advancement, the erosion of privacy, and the automation of work. It's overwhelming, to say the least. But now, it's time to shift gears. Because while the digital landscape might feel like an unstoppable, all-consuming machine, you don't have to be a passive passenger. You have the ability to learn the rules, master the system, and, dare I say it, even turn it to your advantage.

Now it is about strategy. It's about taking control of your digital future before an algorithm does it for you. We'll explore practical ways to prepare for what's coming, from acquiring the skills that will truly matter in the future to maintaining your autonomy in a world increasingly driven by AI. We'll also discuss how to preserve the qualities that make us uniquely human—creativity, empathy, and individuality—in a reality that often prioritizes efficiency over humanity.

The goal here isn't just survival; it's thriving. Let's face it: running away from technology isn't a viable option. It's woven into the fabric of our lives, and its influence will only grow. The only way forward is to embrace it, understand it, and ensure it's working for you, not against you.

So, let's get started. It's time to move from "Help! My phone knows more about me than I do!" to "I am the master of my digital domain." The future is coming, and with the right mindset and tools, you can meet it head-on, not as a victim, but as a participant shaping your own destiny.

Strategy #1: Take Back Control – Prepare for the Future

The digital age is advancing at an unprecedented pace, and with it comes a pressing question: How do we maintain control over our own lives as technology increasingly takes the wheel? Right now, it often feels like we're passive passengers in a high-speed vehicle, with our devices predicting our every move, feeding us algorithmically curated content, and making decisions before we even realize we had a choice. But the truth is, control isn't lost; it's being outsourced. And it's time to reclaim it.

The future doesn't have to be a dystopian nightmare where humans are controlled by machines. By understanding the systems that shape our digital lives and taking proactive steps to manage them, we can ensure that technology serves us, rather than the other way around. This section is about empowering you to take back control, equipping you with the knowledge and tools to navigate the digital landscape with confidence and autonomy.

Tip #1

Become an Active Digital Citizen

One of the most common mistakes people make is assuming they are merely users of technology, rather than active participants in the digital ecosystem. This passive mindset leaves individuals vulnerable to manipulation by algorithms, corporations, and even governments. The first step toward thriving in the modern digital age is shifting from passive consumption to active engagement.

➢ Understand How AI & Algorithms Work

Knowledge is power, and this is especially true in the digital world. The more you understand how platforms operate, the less likely you are to be manipulated by them. Dive into the basics of how recommendation engines shape your social media feed, how your data is collected and monetized, and why companies prioritize engagement over your well-being. For example, social media platforms use algorithms to keep you scrolling by showing content that triggers emotional reactions, often at the expense of your mental health.[88] By understanding these mechanisms, you can make more informed choices about how and when you engage with technology.

➢ Choose Your Digital Diet Wisely

Just as you wouldn't eat junk food for every meal, you shouldn't let your mind consume a nonstop stream of low-quality, algorithm-driven content. Be intentional about what you read, watch, and engage with online. Curate your feeds to prioritize meaningful, educational, and uplifting content. Unfollow accounts that leave you feeling drained or anxious, and seek out sources that enrich your life.

➢ Audit Your Digital Presence

Your digital footprint is larger than you think. Regularly review which apps have access to your data, check your privacy settings, and be mindful of what information you're giving away for convenience. For example, many apps request access to your location, contacts, and camera, often unnecessarily. By tightening your digital security, you can reduce the amount of personal data available to corporations and hackers.

Tip #2
Make Technology Work for You

Technology should be a tool that enhances your life, not a system that dictates your behavior. Developing healthy digital habits is essential for staying in control and maintaining your well-being in an increasingly connected world.

- **Set Boundaries with Your Devices**

Your phone, laptop, and other devices are designed to demand your attention 24/7. But you don't have to give it to them. Use screen time limits, notification management tools, and digital detox practices to create balance in your daily life. For example, designate specific times for checking emails or social media, and turn off notifications during meals, family time, or before bed.

- **Resist the Infinite Scroll Trap**

Social media platforms are engineered to keep you hooked. The infinite scroll feature, autoplay videos, and push notifications are all designed to maximize your time on the app. Training yourself to engage consciously, rather than mindlessly scrolling, can prevent digital exhaustion and help you reclaim your time.

- **Master the Art of Selective Engagement**

Not every email needs an immediate reply, especially if you are only in CC; not every group chat requires your participation, and not every trending topic deserves your outrage. Learn to prioritize your energy and attention. By focusing on what truly matters, you can avoid the burnout that comes from trying to keep up with every digital demand.

Tip #3
Strengthen your Mental Privacy

As technology advances, the concept of privacy is evolving in ways we never imagined. Once upon a time, privacy concerns revolved around protecting your browsing history or keeping your social media accounts secure. But with the rise of BCIs, we're entering an era where even your thoughts might not be entirely private.[89] Governments and corporations may attempt to regulate or access neural data for everything from marketing to security surveillance.

➢ Be Aware of the Mind-Tech Connection

BCIs promise incredible convenience, such as controlling devices with your thoughts or uploading knowledge directly to your brain. But they also introduce significant risks, including the potential for mental privacy invasion. Understanding these risks will help you make informed decisions about adopting new technologies.

So, when the wave of BCI adoption reaches us, be sure to ask the right questions before diving in. Does the device encrypt your neural activity? Can you control what's shared and with whom? Are there laws protecting your cognitive rights? Just as you wouldn't give a stranger full access to your phone, you shouldn't give technology full access to your brain without clear safeguards in place.

Adopting advanced tech should never come at the cost of your autonomy. Stay informed, stay critical, and don't be afraid to say no when a device asks for more than you're ready to give

➢ Defend Cognitive Freedom

Just as encryption protects your digital data, future privacy measures must focus on safeguarding your internal thought processes. Advocating for technologies that prioritize mental privacy will be crucial in ensuring that your thoughts remain your own.

➢ Advocate for Mind Encryption Rights

If thoughts become trackable data, new legal protections will be needed to prevent unintentional oversharing or corporate misuse of cognitive insights. Supporting policies that protect mental privacy will be essential in preserving individual autonomy in the digital age.

Tip #4

Recognize Algorithmic Influence

When a digital system optimizes for something, whether it's the news you see, the ads you encounter, or even the friends you interact with online, it's making decisions about your reality. But who decided what should be optimized?

➢ Question Who Benefits

If an AI-driven platform prioritizes engagement, does that mean it's showing you content that is genuinely beneficial, or simply what keeps you glued to the screen? By questioning the motives behind algorithmic decisions, you can better understand how they shape your behavior.

➢ **Override the Defaults**

Most platforms allow some level of customization. Take advantage of settings that let you reduce algorithmic influence, such as opting for chronological feeds or adjusting ad preferences. By taking control of your digital environment, you can create a more authentic and meaningful online experience.

➢ **Develop Digital Skepticism**

Not every digital convenience is harmless. If a system is making choices for you, ask yourself, is this optimizing for me or for them? Cultivating a healthy skepticism toward technology will help you navigate the digital world with greater awareness and autonomy.

Strategy #2: Acquire Future-Proof Skills – What to Learn and How to Adapt

The future won't belong to those who resist technology but to those who learn to navigate, master, and even thrive alongside it. The pace of change is accelerating at an unprecedented rate, and the skills that kept people employed and relevant a decade ago might be obsolete tomorrow. In this new era, lifelong learning isn't just a nice-to-have; it's a necessity for survival. The question isn't whether you'll need to adapt but how quickly and effectively you can do so.

The good news? The tools and resources to future-proof yourself are more accessible than ever. By embracing a mindset of continuous growth, focusing on skills that complement rather than compete with technology, and using AI as a learning partner, you can position yourself not just to survive but to thrive in the digital age.

Tip #5

Be a Continuous Learner

In a world where change is the only constant, the ability to learn, unlearn, and relearn will be your greatest asset. The days of relying on a single degree or credential to carry you through your career are over. Instead, the future belongs to those who can adapt, evolve, and grow alongside the shifting demands of the digital landscape.

➢ **Embrace a Growth Mindset**

A growth mindset—the belief that abilities and intelligence can be developed through effort and learning—is the foundation of future-proofing yourself. This means viewing challenges as opportunities to grow, embracing failure as a stepping stone to success, and staying open to new ways of thinking. The most successful individuals of the future won't be those who know everything, but those who are willing to learn anything. Those who will continuously live outside their comfort zone and strive to be in the growth zone.

➢ **Stay Curious**

Curiosity is the engine of innovation and adaptability, the spark that drives us to explore, experiment, and evolve. In a world where new technologies, tools, and ideas pop up faster than memes on the internet, staying curious isn't just helpful, it's essential. Ask questions, dive into unfamiliar topics, and chase opportunities to expand your knowledge like a cat chasing a laser pointer. Sure, they say "curiosity killed the cat," but let's not forget the second half of that phrase: "satisfaction brought it back." The professionals who thrive in this fast-paced world will be the ones who never stop asking, "What's next?" because in the digital age, standing still is the real danger. So, stay curious, stay hungry for information, and remember: even if curiosity kills the cat, it's the curious humans who end up inventing the next big thing.

➢ **Leverage AI as a Learning Partner**

AI isn't just a tool for automation; it's also a powerful ally in education, capable of transforming how we learn and grow. AI-driven platforms can personalize learning experiences, tailoring content to your unique strengths, weaknesses, and goals.[90] From language learning apps that adapt to your pace to coding bootcamps that pinpoint exactly where you need help, AI can help you acquire new skills faster and more efficiently than ever before. But here's the catch: while AI can guide your learning, it's crucial to stay in control. The key is to use these tools strategically, ensuring they're doing what you need, not just what the AI thinks you need. After all, AI might be smart, but it doesn't know your aspirations, your passions, or your long-term vision. Use it to complement your learning, not replace your critical thinking. Stay in the driver's seat and let AI serve as the GPS, offering guidance but never deciding the destination.

Tip #6
Master Digital Future Essential Skills

While technical skills are important, they're only part of the equation. The future workforce will require a blend of technical expertise, cognitive abilities, and uniquely human qualities that machines can't replicate. Here are the skills that will set you apart in the digital age:

➢ Human-AI Collaboration

As AI becomes more integrated into the workplace, the ability to work alongside it will be one of the most critical skills of the 21st century. This isn't just about knowing how to use AI tools; it's about understanding how they operate, recognizing their limitations, and leveraging them to enhance, rather than replace, human decision-making. Think of it as a partnership: AI brings speed, precision, and data-driven insights, while humans bring creativity, intuition, and ethical judgment.

To master this skill, you'll need to develop a deep understanding of how AI systems function. This means learning the basics of how algorithms process data, make predictions, and generate recommendations. But it also means recognizing where AI falls short. For example, AI excels at analyzing vast amounts of data and identifying patterns, but it struggles with context, nuance, and the unpredictable nature of human behavior. It can't understand sarcasm, read between the lines, or make value-based decisions. That's where you come in.

Take data analysis as an example. AI can scan through mountains of information and highlight trends or anomalies, but it's up to humans to interpret what those findings mean. Does a spike in sales reflect a successful marketing campaign, or is it a seasonal trend? Is a drop in productivity a sign of burnout, or is it due to external factors like a global crisis? These are questions that require human insight, empathy, and critical thinking, at least for the next decade, but this might change a bit after that in the decades to follow.

Another key aspect of working alongside AI is knowing when to trust its recommendations and when to override them. AI systems are only as good as the data they're trained on, and biased or incomplete data can lead to flawed conclusions. For instance, an AI hiring tool might favor candidates from certain backgrounds because of historical biases in its training data. It's up to humans to spot these biases, question the outputs, and ensure decisions are fair and ethical.

In short, the skill of working alongside AI is about balance. It's about knowing when to let AI handle the heavy lifting and when to step in with human judgment. It's about using AI as a tool to amplify your capabilities, not as a replacement for your unique strengths. By mastering this skill, you'll not only stay relevant in the workplace; you'll thrive in a future where human-AI collaboration is the norm. So, start learning now, because the future isn't just coming; it's already here.

➢ **Critical Thinking & Information Filtering**

In an age of information overload and widespread misinformation, the ability to think critically and filter out noise isn't just important; it's absolutely essential, and it's a skill you must acquire today. This means verifying sources, detecting bias, and questioning assumptions, skills that echo the philosophy of Averroes, the 12th-century Andalusian philosopher who championed reason, logic, and the relentless pursuit of truth. Averroes believed that critical thinking and rational inquiry were the keys to understanding the world and separating fact from fiction. His emphasis on reason as a pathway to knowledge couldn't be more relevant in today's digital landscape, where conflicting information and hidden agendas are everywhere.

Averroes argued that truth could only be uncovered through rigorous analysis and the application of logic, a principle that holds immense value in our modern world. Whether you're evaluating a news article, analyzing data, or making a business decision, critical thinking acts as your shield against manipulation and misinformation.[91] Just as Averroes tried to find balance between belief and logical thinking, we must reconcile the overwhelming flood of information with our ability to discern what's credible and what's not.

To acquire this skill today, start by verifying sources before accepting any information as truth. Check the credibility of the source and look for multiple independent confirmations of the same facts. Detect bias by asking yourself who benefits from the information being presented and whether certain perspectives are being excluded. Question assumptions by digging deeper into why the information is being shared and what evidence supports it. Practice skepticism, especially with sensational headlines and viral content; if it seems too outrageous to be true, it probably is. Engage in active learning by reading widely, exploring opposing viewpoints, and challenging your own beliefs. The more you expose yourself to diverse perspectives, the sharper your critical thinking will become. Finally, use tools to your advantage, such as fact-checking websites, reverse image searches, and AI-powered tools, to verify information quickly and efficiently.

The digital age demands that we think critically, act deliberately, and question relentlessly. By embracing the principles of reason and logic that Averroes championed, we can navigate the chaos of information overload with confidence and clarity. Don't wait; start augmenting your critical thinking skills today, because in a world drowning in data, the ability to think clearly is your lifeline.

- **The AI-Proof Toolkit: Empathy, Leadership, and Creative Problem-Solving**

While AI excels at processing data and performing repetitive tasks with unmatched speed and precision, it struggles with the nuances of human emotion and creativity, areas where humans have an undeniable edge. Skills like empathy, leadership, and creative problem-solving are not just valuable; they are irreplaceable. These uniquely human abilities will remain in high demand, especially as AI takes over more routine and technical aspects of work. Whether you're managing a team, designing a product, or building relationships, your capacity to connect with others on a deeply human level and think outside the box will set you apart in a world increasingly dominated by machines.

Empathy, for instance, is the cornerstone of effective communication and collaboration. It allows you to understand and respond to the emotions, needs, and perspectives of others, whether they're colleagues, clients, or customers. In a workplace where AI handles data and logistics, your ability to foster trust, resolve conflicts, and inspire others will be invaluable. Leadership, too, is a skill that AI cannot replicate. True leadership involves more than just making decisions; it's about motivating others, navigating uncertainty, and creating a vision that others want to follow. While AI can provide data-driven insights, it's humans who inspire action and drive change.

Creative problem-solving is another area where humans shine. AI can analyze patterns and suggest solutions based on existing data, but it lacks the ability to imagine entirely new possibilities or approach problems from unconventional angles. Creativity is what fuels innovation, whether you're designing a groundbreaking product, crafting a compelling marketing campaign, or finding a novel way to streamline operations. It's the spark that turns challenges into opportunities and ideas into reality.

To develop these skills, start by actively practicing empathy in your daily interactions. Listen more than you speak, ask open-ended questions, and strive to understand the emotions and motivations behind people's actions. For leadership, focus on building emotional intelligence, perfecting your decision-making skills, and learning how to inspire and empower others. And for creativity, make time for brainstorming, explore diverse perspectives, and embrace experimentation, even if it means failing a few times along the way.

In a future where AI handles the "what" and the "how," it will be up to humans to provide the "why." By cultivating empathy, leadership, and creativity, you'll not only future-proof your career but also ensure that technology serves you, not the other way around. These skills are your superpower in the digital age; use them to connect, innovate, and lead in ways that machines never can.

➢ **Cybersecurity Awareness**

As our lives become increasingly digital, understanding cybersecurity is no longer optional; it's a necessity. From protecting personal data to safeguarding sensitive information, cybersecurity awareness is a critical skill for individuals. In a world where cyber threats are constantly evolving, being proactive about security can mean the difference between staying safe and falling victim to a devastating breach.

This starts with mastering basic security practices, such as using strong, unique passwords for every account and enabling two-factor authentication (2FA) wherever possible. These simple steps can significantly reduce the risk of unauthorized access to your accounts. But cybersecurity goes far beyond the basics. It also involves recognizing and responding to more sophisticated threats, such as phishing attacks. Phishing, where attackers trick you into revealing sensitive information like passwords or credit card numbers, is one of the most common and dangerous cyber threats. Learning to spot phishing attempts, whether they come through email, text messages, or even social media, is essential. Look for red flags like suspicious sender addresses, urgent or threatening language, and requests for personal information.

Another critical aspect of cybersecurity is staying informed about emerging threats. Cybercriminals are constantly developing new tactics, from ransomware that locks your files until you pay a ransom to social engineering schemes that manipulate you into giving up sensitive information. Staying updated on the latest threats and understanding how they work can help you avoid falling victim. For example, knowing how to identify a fake website or a malicious attachment can save you from malware infections or identity theft.

You must be cautious about the apps you download, the links you click, and the networks you connect to. Public Wi-Fi, for instance, can be a hotspot for cyberattacks, so using a virtual private network (VPN) to encrypt your connection is a smart move.

Cybersecurity also extends to protecting your digital footprint. This means being mindful of what you share online, adjusting privacy settings on social media, and regularly reviewing which apps and services have access to your data. Even seemingly harmless information, like your birthday or pet's name, can be used by hackers to guess passwords or answer security questions.

In short, cybersecurity awareness is about more than just avoiding risks; it's about taking control of your digital life. By understanding the threats, adopting best practices, and staying informed, you can protect yourself and your data in an increasingly connected world. Whether you're an individual or part of an organization, cybersecurity is a skill that empowers you to navigate the digital landscape with confidence and peace of mind.

Strategy #3: Navigate the Moral Maze of AI wisely

The rise of AI, automation, and algorithmic decision-making is dragging us into a labyrinth of ethical and existential dilemmas that would make even a sci-fi writer sweat. Who gets to decide how AI is programmed? How do we ensure fairness when algorithms are already biased? What happens when AI knows us better than we know ourselves? And the million-dollar question: What does it even mean to be human in a world that's increasingly optimized for machines?

These aren't just thought experiments for late-night philosophy debates; they're urgent realities we need to tackle today. The way we answer these questions will shape the world we live in tomorrow. So, buckle up, because this section is your survival guide to navigating the moral minefield of the digital age. We'll explore the biggest ethical challenges, from accountability to free will, and arm you with the tools to keep your humanity intact in a world obsessed with efficiency.

Tip #7

Always Ask Who's Accountable When AI Gets It Wrong

AI is already making high-stakes decisions, approving loans, diagnosing diseases, influencing hiring, and even assisting in legal rulings. But when things go sideways, who takes the fall? If an AI-driven car crashes, is it the fault of the manufacturer, the programmer, or the poor soul behind the wheel? If an algorithm unfairly denies someone a job, who gets the blame?

Many AI systems are "black boxes," meaning even their creators can't fully explain how they make decisions. That's why we need to demand transparency. AI should be able to explain its reasoning in plain English (or whatever language you speak). Governments and regulatory bodies also need to step up, creating clear guidelines to ensure AI systems are fair, unbiased, and always have human oversight for critical decisions. And let's not let tech giants off the hook. They need to prioritize ethics over profits and take responsibility for the societal impact of their creations.

Tip #8

Stay Aware of How Tech Nudges Your Decisions

AI is getting scarily good at predicting human behavior. It knows what you want before you do, nudging you to buy that pair of shoes, read that article, or swipe right on that profile. But if AI is steering your decisions, are they really your decisions?

The first step to reclaiming your free will is awareness. Understand how recommendation algorithms work; they're not magic, just math designed to keep you hooked. Next, practice digital minimalism. Cut back on algorithmically curated content and seek out diverse perspectives. And let's not forget advocacy: as AI shapes more of our choices, we need policies that protect our digital autonomy and ensure we're not just puppets in an algorithmic play.

Tip #9

Guard Your Mental Space Your Thoughts Are the Final Frontier

With advancements in BCIs, we're entering an era where technology might one day read or even manipulate your thoughts. Sounds cool until you realize your brainwaves could become the next hot commodity for corporations and governments.

The battle for cognitive freedom is already heating up. If thoughts become data, we'll need new privacy laws to protect them. Think of it as mind encryption; your thoughts should be as private as your medical records. And let's not even get started on the dystopian nightmare of thought policing. Defending mental privacy isn't just a personal responsibility; it's a societal priority.

Strategy #4: Reclaim Meaning in a Digital World by Defining Your Own Purpose

In a world where work is increasingly automated, relationships are mediated by screens, and artificial intelligence makes decisions before we even realize there was a choice to be made, the biggest challenge we face is not just how to survive but how to live with purpose. The digital age offers unprecedented convenience and connectivity, yet it also threatens to strip away the fundamental human experiences that give life meaning. When AI can generate art, automate careers, and optimize our daily decisions, what remains uniquely ours? What gives our lives depth, fulfillment, and authenticity?

Reclaiming meaning in the digital world is not about rejecting technology; it's about actively defining what matters to us in a landscape where machines and algorithms increasingly dictate the terms of existence. Now we explore how to carve out a sense of purpose, identity, and fulfillment in a future where traditional definitions of work, relationships, and personal achievement may no longer apply.

Tip #10

Redefine Success Beyond Work

For centuries, work has been a cornerstone of identity and purpose for a significant portion of the global population. It's how we've defined ourselves, contributed to society, and met our basic needs. But as AI and automation continue to disrupt industries and redefine the nature of labor, the role of work in our lives is undergoing a seismic shift. If jobs are no longer central to human survival, thanks to machines handling everything from manufacturing to data analysis, what will give our lives meaning and direction?

The traditional idea that a career is the primary measure of success is quickly becoming outdated. To future-proof your sense of purpose, it's essential to detach your identity from your job title and instead focus on cultivating personal projects, intellectual growth, and creative pursuits that fulfill you beyond financial necessity. With automation taking over repetitive and ordinary tasks, humans will have more freedom to engage in activities that require creativity, emotional intelligence, and community-driven efforts.

This shift means that pursuing work aligned with your personal values, rather than simply working to survive, will become increasingly important.

As AI reshapes economies, new financial models like Universal Basic Income could emerge, giving people the freedom to focus on personal growth, artistic expression, and solving meaningful problems instead of just working relentlessly to pay the bills. Imagine a world where you can finally write that novel, learn to paint, or solve world hunger, all while not worrying about your next paycheck. Sounds dreamy, right? But let's be real: the idea of a single financial model working everywhere feels a bit like expecting a cat to fetch your slippers. When has our planet ever had one functional financial system that worked for everyone? From exchanging goats to trading Bitcoin, humanity's track record with money is more chaotic than a toddler with a crayon. So, while Universal Basic Income might sound like the golden ticket, let's not hold our breath waiting for it to save the day. In the meantime, maybe we can start by not breaking the systems we already have.

If this shift happens, it will have the potential to redefine what success looks like, moving away from climbing the corporate ladder and toward contributing to society in ways that feel personally fulfilling and socially impactful. Success in the future may be less about titles and salaries and more about the legacy you leave behind, the connections you build, and the creativity you bring to the world.

In this new era, the challenge will be to find purpose beyond traditional work. It's about embracing the freedom that automation offers to explore what truly matters to you, whether that's creating art, mentoring others, solving global challenges, deepening your spiritual connection, or simply living a life rich with experiences and relationships. The future of work may be uncertain, but the opportunity to redefine success on your own terms has never been greater.

Tip #11

Maintain Human Connection

Digital convenience has paradoxically made human relationships more superficial than ever. We can message thousands of people in an instant, yet loneliness has reached epidemic levels. No virtual experience can fully replace the nuance, warmth, and spontaneity of face-to-face conversation. Making time for real-world social connections will become a deliberate act of resistance against digital isolation. AI-driven social bots and virtual companions may become so advanced that they feel more emotionally satisfying than real human relationships. While these interactions might be convenient, they lack genuine depth and unpredictability, both of which are essential for meaningful connection.

In a world where attention spans are shrinking, engaging in deep, thoughtful discussions without digital distractions will become a rare but invaluable skill. Fostering emotional intelligence will be key to maintaining meaningful relationships.

Tip #12

Balance your Digital and Physical Reality

As VR, AR, and immersive technologies continue to advance, the temptation to fully transition into digital existence will be stronger than ever. Whether it's the metaverse, hyper-realistic AI-generated environments, or neuro-linked experiences, we must notice that digital immersion disconnects us from real life.

Digital spaces offer endless stimulation and convenience, but at the cost of physical experiences, unpredictability, and personal growth. Establishing a healthy balance between digital engagement and real-world presence will be vital. In a world of infinite entertainment, boredom has become nearly extinct, but boredom is often where creativity, self-reflection, and meaningful insight emerge. Allowing the mind to wander without distraction is an essential practice for maintaining individuality and imagination.

Setting boundaries for digital consumption, whether through device-free meals, nature retreats, or daily mindfulness practices, can help preserve a sense of presence and clarity in an overwhelmingly connected world.

Tip #13

Cultivate Purpose in an AI-Dominated Society

AI lacks empathy, intuition, and moral reasoning, qualities that are exclusively human. Focusing on what makes us uniquely valuable will be key to finding purpose. True fulfillment often comes from solving problems, pushing boundaries, and creating something that matters. Whether through scientific discovery, artistic innovation, social activism, or personal development, the future of human purpose will depend on our ability to constantly evolve and contribute.

The shift from survival-based work to purpose-driven existence will make community engagement, mentorship, and knowledge-sharing more valuable than ever. Building and strengthening communities will be essential for fostering a sense of belonging and shared purpose.

Strategy #5: Build Resilience in the Digital Age

Let's face it: the digital age is like a never-ending rollercoaster that forgot how to stop. Technological advancements are sprinting ahead while we're still trying to figure out how to mute group chat notifications. Information overload has turned decision-making into a game of "Which option will I regret the least?" And the always-online culture? It's like being stuck in a Zoom meeting that never ends. In this whirlwind of change, resilience isn't just a nice-to-have—it's your lifeline.

To thrive in this fast-moving, always-evolving reality, you need more than just a strong Wi-Fi connection. You need mental grit, emotional stability, and practical adaptability. Think of it as your personal toolkit for staying balanced, in control, and ready to tackle whatever the digital world throws at you. So, let's dive in and build that resilience muscle—because the future isn't waiting for you to catch up.

Tip #14
Cultivate Mental Resilience

If your brain feels like a browser with 47 tabs open, you're not alone. Between endless notifications, 24/7 news cycles, and the infinite scroll of doom, mental resilience is your secret weapon for staying grounded. It's about keeping your cool when the digital world feels like it's spinning out of control.

- **Create Digital Boundaries**

Your phone doesn't need to be your third arm—seriously, it's not a limb. Set screen time limits to keep your scrolling in check, mute those non-essential notifications, and carve out tech-free zones in your home. Yes, that means no Instagram in the bathroom. Let's be honest, your brain doesn't need to process memes while you're brushing your teeth.

By creating these boundaries, you're giving your mind the space it needs to breathe, recharge, and remember what it's like to exist without a screen in your face. Trust me, your brain will thank you, and so will your sanity.

- **Practice Mindfulness and Reflection**

In a world that thrives on urgency, where every ping, buzz, and notification demands your immediate attention, the ability to pause is nothing short of a superpower. It's like hitting the brakes in a car that's been speeding nonstop. Whether it's through meditation, deep breathing exercises, prayer, or just staring at a wall for five minutes, mindfulness helps you reset and refocus. It's your mental Ctrl+Alt+Delete, a way to clear the clutter in your brain and start fresh. Taking these moments of stillness isn't just a luxury; it's a necessity for staying grounded in a world that's always rushing. So, give yourself permission to pause. Your mind will thank you, and you'll come back sharper, calmer, and ready to tackle whatever comes next.

- **Prioritize Rest and Recovery**

Rest isn't laziness; it's strategy. Think of it as your secret weapon for staying sharp, creative, and productive in a world that never seems to slow down. Step away from the screen, take a walk in nature, or dive into a hobby that doesn't involve Wi-Fi, whether it's painting, gardening, or just daydreaming on the couch. A well-rested mind is like a well-oiled machine: it runs smoother, lasts longer, and doesn't crash as often. When you give yourself permission to rest, you're not just recharging your batteries; you're setting yourself up to perform at your best. So don't feel guilty about taking a break. It's not downtime; it's upgrade time. Your future self will thank you for it.

Tip #15
Strengthen Emotional Resilience

Here's the irony of the digital age: we're more connected than ever, yet loneliness is at an all-time high. Emojis are great, but they're no substitute for a real hug or a face-to-face conversation. Emotional resilience is about staying connected to what truly matters even when the world feels virtual.

- **Prioritize Real-World Connections**

Put down the phone and pick up a coffee cup or a tea mug if that's more your vibe. Make time for family dinners where the only screens are the ones on the microwave. Meet friends for lunch and actually look at their faces instead of your notifications. Join a community group, whether it's a book club, a yoga class, or a local volunteer organization. Real-world interactions are like emotional vitamins. They nourish your soul, boost your mood, and keep you grounded in what truly matters. In a world where so much of our connection happens through screens, these face-to-face moments are a reminder that life is richer when it's shared. So, step away from the digital noise and into the warmth of real human connection. Your heart and your mental health will thank you.

- **Curate a Healthy Digital Environment**

Not all digital spaces are the same. If your social media feed feels like a mess filled with negativity, drama, or constant noise, it's time to clean it up. Unfollow accounts that drain your energy, spread toxic messages, or make you feel worse about yourself. Instead, fill your feed with content that inspires, teaches, or brings you joy. Follow people and pages that motivate you, share useful knowledge, or simply make you smile. Your digital space should support your well-being, not add to your stress. You have the power to choose what you see online, so choose wisely. A positive, intentional feed isn't just a bonus; it's an important part of staying mentally healthy.

- **Recognize When to Seek Support**

Emotional resilience doesn't mean going it alone. If you're feeling overwhelmed, talk to a therapist, join a support group, or chat with a trusted friend. Asking for help isn't a sign of weakness; it's a sign of strength.

Tip #16

Develop Practical Resilience

The digital economy is like a game of musical chairs, except the chairs are jobs, and the music never stops. Practical resilience is about staying prepared, adaptable, and ready to pivot when the ground shifts beneath your feet.

- **Diversify Income Streams**

Relying on a single source of income in the age of automation is like putting all your eggs in one basket and then handing that basket to a robot. Let's be real: robots do not care about your financial security. As AI and automation continue to reshape industries, jobs that once seemed stable are becoming vulnerable to disruption. The days of working one job for 40 years and retiring with a gold watch are long gone. The future belongs to those who embrace financial flexibility and diversify their income streams.

Exploring freelancing, side hustles, or investments isn't just a smart move; it is a survival strategy. Freelancing allows you to leverage your skills on your own terms, whether it is writing, designing, consulting, or teaching. Side hustles like starting a small business or selling handmade goods can provide an extra cushion while also giving you a creative outlet. And investments, whether in stocks, real estate, or even cryptocurrency, can help grow your wealth over time, creating a safety net for the future.

But it is not just about the money. Diversifying your income also gives you freedom and control over your life. If one income stream dries up, you have others to fall back on. It is like building a financial safety net around your home. You are protecting yourself from the unpredictability of the digital economy. Plus, having multiple income sources can open doors to new opportunities, skills, and experiences that you might not have considered otherwise.

The key is to start small and stay consistent. You do not need to quit your day job tomorrow or become a Wall Street guru overnight. Begin by identifying your skills, passions, and resources, and then explore ways to monetize them. Maybe it is turning a hobby into a side job or investing a small amount each month into a diversified portfolio. The goal is not to become a millionaire overnight. It is to create options for yourself so you are not left scrambling when the robots come for your job.

In a world where change is the only constant, financial flexibility is your best defense. It is about taking control of your future, one egg at a time, and making sure no single basket or robot can knock you off course. So, start diversifying today. Your future self will thank you.

- **Build a Professional Network**

Your network is your safety net. In a world where industries shift overnight, jobs evolve faster than ever, and opportunities often come from unexpected places, the people you know can make all the difference. Cultivating relationships with mentors, collaborators, and peers isn't just a nice-to-have; it's a critical part of thriving in the digital age.

Mentors are like the GPS of your career; they've been where you're going and can help you navigate the twists and turns. They offer wisdom, guidance, and a reality check when you need it most. Collaborators, on the other hand, are your co-pilots. They bring fresh perspectives, complementary skills, and the kind of creative energy that turns good ideas into great ones. And peers? They're your cheerleaders, sounding boards, and sometimes even your competition, pushing you to be better while reminding you that you're not alone in the journey.

But building a strong network isn't just about collecting business cards or LinkedIn connections. It's about fostering genuine relationships. Show up for others, offer help when you can, and don't be afraid to ask for advice or support when you need it. Networking is a two-way street, and the best relationships are built on mutual respect and shared value.

Remember, it's not just what you know, it's who you know. A strong network can open doors to opportunities you didn't even know existed. It can provide a lifeline during tough times, a sounding board for big decisions, and a source of inspiration when you're feeling stuck. In a world where automation and AI are reshaping the workforce, human connections are more valuable than ever.

So, invest in your network. Attend industry events, join professional groups, and don't shy away from reaching out to someone you admire. The relationships you build today could be the key to your success tomorrow. After all, your network can be more than just a safety net. It could also be the launchpad for bigger and better opportunities.

Tip #17
Embrace Change as a Constant

Change is the only constant in the digital age, and the most resilient people aren't the ones who resist it; they're the ones who embrace it. Think of change as your personal trainer: it's tough, but it's making you stronger.

➢ **Stay Open to New Experiences**

The future will bring opportunities that don't even exist yet. Jobs, technologies, and industries that seem like science fiction today could be mainstream tomorrow. Staying open to new experiences is your ticket to staying relevant in this ever-changing landscape.

Curiosity is your greatest ally. Explore new skills, dive into unfamiliar topics, and don't be afraid to step outside your comfort zone. Whether it's learning to code, trying your hand at content creation, or exploring a completely different career path, every new experience adds to your toolkit and prepares you for the unknown.

And when the time is right, don't hesitate to pivot. Pivoting isn't a sign of failure; it's a sign of growth. Maybe your current role no longer excites you, or maybe a new opportunity aligns better with your passions and values. Whatever the reason, embracing change and taking calculated risks can lead to unexpected and rewarding paths.

➢ **Recognize that Resilience is Built, Not Given**

No one is born resilient; it's a skill you develop through experience, trial, and error. Think of resilience as a muscle: the more you use it, the stronger it gets. Every setback, every challenge, and every failure is an opportunity to learn, adapt, and come back stronger.

Resilience isn't about avoiding difficulties; it's about facing them head-on and finding a way forward. It's about learning from your mistakes, adjusting your approach, and refusing to let obstacles define you. Each time you overcome a challenge, you're not just surviving; you're building the resilience you'll need for the next one.

And remember, resilience isn't a solo journey. Lean on your support network, seek advice from mentors, and don't be afraid to ask for help when you need it. Resilience isn't about being invincible; it's about being resourceful.

Final Reflection: Your Role in the Digital Future

The future is not a distant storm we must brace for. It is already here, swirling around us, shaping how we live, communicate, and make meaning. The digital age is not just changing the tools we use; it is transforming who we are. But amidst all the change, one thing remains constant: your power to choose how you respond.

You are not just a user. You are not merely a consumer of content, a number in someone's engagement metrics, or an employee working for a system you don't understand. You are a human being with the power to choose and shape your own path. That is your greatest asset in this era of algorithms and automation.

The pace of technology will continue to accelerate. Innovations in AI, BCIs, and immersive digital environments will radically reshape daily life. The temptation will be to let the convenience wash over you, to accept every new update as inevitable, to automate decisions, outsource memory, and allow systems to optimize your existence for you. But if you do, you risk forgetting what it means to be human.

Now is the moment to slow down and act deliberately. You are not here to keep up with machines; you are here to lead, to imagine, to feel, and to connect. That starts with reclaiming control over the digital tools and platforms you interact with every day. Learn how they work. Understand what drives their decisions. Make conscious choices about what you consume, who you follow, how you use your time, and what you say yes to.

The more you build your digital literacy, the less susceptible you become to manipulation. Set boundaries. Turn off the endless notifications. Say no to the infinite scroll. Choose stillness over noise, clarity over overwhelm. These are not just digital habits; they are acts of resistance. They are how you defend your autonomy.

But survival in the digital age requires more than resisting distraction. It requires learning constantly and intentionally. In a world where information is abundant and change is relentless; learning becomes your superpower. And not just learning in the traditional sense, learning how to think, how to evaluate, and how to stay curious. The mindset you carry into each new decade will shape not only your success but also your well-being.

Focus on building future-proof skills. Some of these will be technical, but many will be deeply human. Machines can predict, calculate, and execute. But they cannot lead with empathy. They cannot dream, love, or seek justice. They cannot take a moral stand. That's your domain. Emotional intelligence, creative thinking, and ethical reasoning—these are the muscles you must build, because these are the capabilities that machines will never master.

Remember that your work does not define your worth. As automation reshapes the job market, you will need to redefine success. Perhaps your next chapter won't be measured by promotions, paychecks, or productivity metrics. Perhaps it will be measured by the relationships you cultivate, the art you create, the ideas you pursue, the wisdom you pass on, or the care you offer others. This is not regression; it's evolution.

As society transitions toward a future where work may no longer be central to our identity, you will be invited to ask deeper questions. Who are you without your title? What gives your life purpose? These are the questions machines cannot answer for you. Only you can.

In this new era, spiritual and emotional grounding will matter more than ever. Whether through faith, mindfulness, community service, or reflection, find a compass that keeps you centered in a world that will often try to spin you in circles. Connection—real connection—will be your anchor. Prioritize it. Invest in it. Don't let digital interaction replace eye contact, shared meals, or deep conversations. They are irreplaceable.

As we look forward, your digital self and your physical self will no longer be separate. They will blur. Who you are online will increasingly shape how people see you, what opportunities come your way, and how you understand yourself. That makes it even more important to be intentional about the life you're building in both spaces. Make sure they reflect your values, your goals, and your humanity.

Let those tips be your turning point. Not the moment you unplugged entirely, but the moment you stepped into the driver's seat of your digital life. The moment you decided to master the system instead of being mastered by it.

You don't need to be a tech expert to do this. You need to be awake. You need to care. You need to ask questions. You need to hold your ground. Technology is not destiny. The future is not fixed. It is being written in every line of code, in every policy debate, and in every individual decision, including yours.

You may not be able to stop the tide of change, but you can learn how to surf. You can build a life that uses technology without being used by it. You can design your habits, your learning, your relationships, and your legacy with intention. You can be the kind of human this digital future needs, one who leads with heart, who questions deeply, who resists the easy path, and who always chooses meaning over momentum.

So, take a moment. Reflect on where you are and where you want to go. Reimagine your relationship with your devices, your data, and your time. Decide what kind of digital citizen you want to be. And then, act.

Because the future will not be determined by the most advanced machine. It will be shaped by the most courageous minds.

Stay mindful. Stay human. And above all, stay in the game, not as a follower of the future but as one of its architects.

Chapter 7
The Digital Bubble and the Family - Navigating Challenges and Thriving in the Digital Age

By 2025, the digital bubble has become an inescapable reality for families worldwide. What began as a convenient tool for communication and entertainment has evolved into an all-encompassing force, shaping how families interact, learn, and grow. Screens are no longer just devices; they are gateways to virtual classrooms, remote workspaces, and social arenas. Artificial intelligence has become a silent partner in parenting, offering everything from personalized learning for children to automated household management. Yet, as the digital bubble expands, so do its challenges. Families are struggling with screen addiction, the blurring of boundaries between work and home life, and the erosion of face-to-face connections. The question is no longer whether technology will influence family life, but how families can adapt to ensure it enriches rather than diminishes their well-being.

We will now dive into the challenges the digital bubble imposes on families in 2025 and explore the emerging issues that will define the next three decades. From the rise of immersive virtual realities to the ethical dilemmas posed by AI-driven decision-making, families must prepare for a future where the line between the physical and digital worlds becomes increasingly blurred. It is a call to action. Families have the power to shape their relationship with technology, ensuring it aligns with their values and strengthens their bonds.

This is a practical guide for families to not only survive but thrive in the digital age. It provides actionable strategies for parents, caregivers, and children to build digital resilience, foster meaningful connections, and create a home environment where technology serves as a tool rather than a tyrant. From setting healthy screen time boundaries to navigating the complexities of online safety, this is a roadmap for families to take control of their digital lives. It also explores how extended families can stay connected across distances, leveraging technology to bridge gaps while preserving the warmth of real-world relationships.

The digital bubble is here to stay, but families don't have to be passive participants in its evolution. By understanding the challenges and embracing practical solutions, families can ensure that technology enhances their lives rather than undermining them. This is your guide to building a future where the digital bubble becomes a source of strength, connection, and growth for your family.

Part 1: Today's Family Challenges

What was once a source of convenience, instant communication, endless knowledge, and limitless entertainment has now become a double-edged sword. Families today are navigating an entirely new landscape, one that demands vigilance, adaptability, and intentionality. Without clear strategies and proactive engagement, many will struggle to maintain strong relationships, safeguard their children's well-being, and prepare for the future.

The reality is hard: the digital world is no longer just a tool or an extension of life. It is a parallel universe, an ever-expanding metropolis that families are forced to exist in, whether they are ready for it or not. And just as no responsible parent would allow their child to wander alone through a vast, unpredictable city without guidance or protection, no child should be left to navigate the digital world unaccompanied. A police officer specializing in cybercrimes against children put it bluntly: 'The internet is a city with no borders, no curfews, and no guaranteed protection. You wouldn't let your child roam the streets of an unknown metropolis alone, so why would you allow them to do so in the digital world?' This comparison highlights the urgent need for parents to rethink their role, not as passive bystanders, but as active guides in their children's digital lives.

Challenge #1: The Rise of Digital Dependency: A Fragmented Family Reality

The digital bubble has fundamentally reshaped family dynamics. While it promises connection and convenience, it often delivers the opposite: an illusion of closeness that masks a reality of disconnection. Families may physically share the same space, but within the digital bubble, each member is often isolated in their own virtual world.[92] The living room, once a place of shared stories and laughter, can turn into a group of individuals absorbed in their screens, disconnected from those sitting right beside them.[93] This phenomenon, often described as "alone together," highlights how the digital bubble can gradually erode the very bonds it was meant to strengthen.

For children, the digital bubble serves both as a playground and a prison. Growing up in a screen-first environment, their earliest interactions with the world are increasingly filtered through digital devices. The natural desire to explore outdoors, engage in imaginative play, or build face-to-face connections with peers is frequently replaced by a reliance on tablets, smartphones, and gaming consoles. These tools provide access to learning and entertainment, but they can also hinder the development of social and emotional skills that flourish through real-world experiences. Within this bubble, children risk losing touch with the hands-on, unpredictable, and deeply human aspects of life.

Social media, a central component of the digital bubble, adds another layer of complexity. For children and teenagers, the process of developing identity is now shaped more by curated online personas than by authentic interactions. The pressure to present an idealized version of oneself, optimized for likes and shares, can distort self-image and interfere with the formation of a grounded and confident identity.[94] This constant digital performance often replaces meaningful relationships, leaving many young people feeling anxious, isolated, and unsure of who they truly are.[95] In this way, the digital bubble becomes a maze of reflections that can be difficult to navigate.

Parents, too, are deeply affected. As discussed earlier in this book, the line between work and home has become increasingly blurred. Smartphones and laptops bring professional responsibilities into family spaces, making it difficult to be fully present.[96] Social media notifications, breaking news, and endless streams of content constantly vie for attention. Even with the best intentions, parents often find themselves stuck in the same cycle of distraction they hope their children will avoid. The result is a family atmosphere where genuine connection is often sacrificed for the illusion of connectivity.

The long-term consequences are troubling. Families risk becoming emotionally fragmented, with relationships built on shallow exchanges rather than meaningful connection. Children may grow up without the resilience, empathy, or communication skills they need to succeed in the real world. As the digital bubble expands, the very foundation of family life, which depends on trust, shared experiences, and emotional closeness, becomes more vulnerable to erosion.

Still, there is hope. The digital bubble does not have to control family life. By recognizing its influence and making intentional choices, families can reclaim their time and connection.

Challenge #2: The Mental Health Crisis in the Digital Bubble: A Growing Epidemic

The digital bubble has launched an era of unprecedented connectivity, but it has also brought with it a silent crisis: the mental health toll of digital overload. For children and teenagers, whose brains are still in critical stages of development, the consequences are particularly severe. Unlike adults, young people are biologically more vulnerable to the addictive design of digital platforms. The prefrontal cortex—the part of the brain responsible for impulse control, decision-making, and self-regulation—is not fully developed until the mid-20s.[97] This makes it incredibly difficult for children and teenagers to resist the powerful engagement mechanisms embedded in games, social media, and streaming platforms. Their brains aren't built to handle the nonstop flood of attention-grabbing content.

The mental health impact of this digital overload is staggering. Anxiety and depression rates among teenagers have surged in recent years, with studies linking this rise to excessive screen time and social media use.[98] For example, a 2023 study published in the *Journal of Adolescent Health* found that teenagers who spend more than three hours a day on social media are twice as likely to report symptoms of depression compared to their peers who use it less.[99] Social media fosters an always-on culture, where self-worth is increasingly tied to online validation. Likes, comments, and followers have become metrics of acceptance, creating a toxic environment of social comparison. Teenagers are constantly measuring themselves against curated, idealized versions of their peers, leading to feelings of inadequacy, low self-esteem, and even body dysmorphia. Real-world cases, such as the tragic story of 14-year-old Molly Russell in the UK, who took her own life after being exposed to harmful content on Instagram, highlight the devastating consequences of unregulated digital consumption.

Excessive screen time is also impairing cognitive development in profound ways. Research from the *American Academy of Pediatrics* indicates that constant digital engagement reduces attention spans, weakens problem-solving abilities, and lowers emotional intelligence.[100] For instance, a 2022 study from the *University of California* found that children who spent more than two hours a day on screens scored significantly lower on language and thinking tests compared to those with limited screen time.[101] The immersive nature of digital content, whether it's fast-paced video games or endless social media scrolling, rewires the brain to crave immediate gratification. This makes it harder for children to tolerate boredom, engage in deep thinking, or persevere through challenging tasks. Over time, this could result in a generation less capable of long-term planning, critical thinking, and emotional regulation. While those are exactly the capabilities they need to thrive in the next decades.

The pressure to stay connected is another significant source of stress for young people. Many children and teenagers feel an obligation to respond to every message, notification, or social media update, fearing social exclusion if they don't. This "fear of missing out" creates a relentless cycle of anxiety and exhaustion. A 2023 survey by *Common Sense Media* revealed that 45% of teenagers feel overwhelmed by the need to be constantly available online.[102] This pressure not only heightens stress levels but also disrupts sleep patterns, as many teenagers stay up late to keep up with their digital lives.[103] Poor sleep, in turn, amplifies mental health issues, creating a vicious cycle that is difficult to break.

Looking ahead, the long-term implications of this digital overload are deeply concerning. If current trends continue, we risk raising a generation that is less equipped to handle the complexities of real-world challenges. The ability to focus, think critically, and regulate emotions—skills essential for personal and professional success—could be significantly diminished. Over the next three decades, this could lead to a society where mental health issues are even more pervasive, and individuals struggle to form meaningful relationships or achieve long-term goals.

Addressing this crisis requires urgent action. Families, educators, and policymakers must work together to create healthier digital environments for young people. We can help children and teenagers navigate the digital bubble in a way that supports their well-being and prepares them for a balanced, fulfilling future. The stakes are high, but with collective effort, we can turn the tide on this growing epidemic.

Challenge #3: The Struggle to Set Boundaries and Manage Digital Consumption

In the digital bubble, parents are waging an uphill battle against some of the most sophisticated attention-capturing technologies ever created. Unlike previous generations, where limiting TV time or enforcing bedtime routines was relatively straightforward, today's parents are up against entire industries engineered to monopolize attention.[104] Apps, games, and streaming platforms are designed the same as casinos with psychological precision, leveraging features like infinite scrolling, autoplay, and streak rewards to keep users, especially children, engaged far beyond healthy limits.[105] These AI-driven engagement tactics are not just tools for entertainment; they are powerful mechanisms that exploit cognitive vulnerabilities, making it incredibly difficult for parents to set and enforce boundaries.

The challenge of managing digital consumption is further complicated by the lack of clear guidelines. Every decision is loaded with uncertainty: How much screen time is too much? Should children have social media accounts? When is the right age for a smartphone? Parents are bombarded with conflicting advice, societal pressures, and the ever-shifting norms of a digitally saturated culture. For instance, while some experts recommend no screen time for children under two, others argue that educational apps can be beneficial. Similarly, the decision to allow a child on social media is often influenced by peer pressure, with many parents feeling compelled to give in because "everyone else is doing it." This ambiguity leaves parents second-guessing their choices and struggling to establish consistent rules.

The rise of remote work and digital lifestyles has further blurred the lines between family time and screen time. For many parents, the expectation to always be available for work emails, video calls, or messages has made it nearly impossible to enforce true family time. The home, once a sanctuary from external demands, has become a space where digital presence often outweighs physical presence. A parent might be physically present at the dinner table but mentally preoccupied with a work notification, while a child scrolls through TikTok beside them. This erosion of boundaries not only undermines family bonding but also sets a poor example for children, who learn to prioritize digital engagement over real-world interactions.

This ongoing struggle is fundamentally reshaping the role of parenting. In the past, parents focused on nurturing values, teaching life skills, and fostering emotional connections. Today, they must also serve as digital gatekeepers, constantly navigating the tension between the benefits of technology and the need for human connection. This dual role requires parents to stay informed about the latest digital trends, understand the psychological tactics used by tech companies, and develop strategies to mitigate their impact. It's a daunting task, especially when many parents themselves are struggling with their own digital dependencies.

The consequences of failing to set effective boundaries are significant. Children who grow up without clear limits on digital consumption are more likely to develop unhealthy habits, such as screen addiction, poor sleep hygiene, and difficulty focusing on offline activities. Over time, this can lead to a diminished capacity for deep thinking, creativity, and emotional regulation. For families, the lack of boundaries can result in fragmented relationships, where digital distractions take precedence over meaningful conversations and shared experiences.

Despite these challenges, there is hope. By adopting proactive strategies, parents can reclaim control over their family's digital lives. Ultimately, the struggle to set boundaries in the digital bubble is not just about limiting screen time; it's about preserving the essence of family life. By prioritizing human connection and offline experiences, parents can create a home environment where technology serves as a tool rather than a tyrant. The journey is not easy, but with intention and effort, families can navigate the digital age in a way that strengthens their bonds and nurtures their well-being.

Challenge #4: Online Safety & Cyber Risks: A Parent's Worst Nightmare

As a parent myself, I urge you to pay close attention if you are a fellow parent or caretaker. This is not just important; it is urgent. Please read carefully. What you are about to learn could be the difference between protecting your child's future and unknowingly leading them into the lion's den of the digital world.

Imagine this: your child is sitting in their room, laughing at a funny video on their phone. They seem safe, happy, and close by. You can hear their giggles through the door, and for a moment, everything feels normal. But in reality, they are not just in their room; they are wandering alone through a lawless digital metropolis. This is the internet in 2025, a place where predators lurk in the shadows, scams unfold in broad daylight, and dangers hide behind every click, every swipe, and every innocent-looking notification. The digital bubble, for all its wonders, is also a breeding ground for threats that can shatter lives in an instant. And your child? They are not merely a user in this world but a target.

This reality keeps me up at night. It should keep you awake too. I know the joy of seeing my child immersed in something they love, whether it's a funny video, a game, or a chat with friends. But I also know the chilling truth: the digital world is not the safe space we want it to be. It's a place where the lines between fun and danger are blurred, where the next click could lead to something harmless or something even horrifying. I've seen the headlines, read the stories, and seen parents who've lived through the unthinkable. It's not just something that happens to "other people's children." It could happen to mine. It could happen to yours.

Cyberbullying is no longer confined to the schoolyard. It follows your child everywhere, through their phone, their tablet, and their gaming console. It invades their bedroom, their dinner table, and their quietest moments. Unlike traditional bullying, which ends when the school bell rings, cyberbullying is relentless. A single cruel comment, a humiliating photo, or a malicious rumor can spread like wildfire, leaving your child with no escape. The psychological toll is devastating: anxiety, depression, and even suicidal thoughts are alarmingly common among victims. And here's the chilling truth: your child might never tell you it's happening. They might feel ashamed, scared, or even blame themselves. As a parent, the thought of my child suffering in silence is unbearable.

The internet is a hunting ground, and predators are the wolves. They don't hide in dark alleys; they hide in plain sight, disguised as friendly gamers, chatroom buddies, or even AI chatbots. Grooming tactics have evolved, becoming more sophisticated and harder to detect. Predators exploit popular platforms such as social media, gaming networks, and even educational apps to gain your child's trust. They manipulate, deceive, and exploit, often without raising suspicion until it's too late. And don't make the mistake of thinking, "It won't happen to my child." Thousands of children fall victim to online exploitation every year, and the numbers are rising. Your child could be next. As a parent, the idea that someone could prey on my child while I'm in the next room is a nightmare I can't shake.

Your child's personal information is a goldmine, and they're giving it away without a second thought. From signing up for harmless-seeming apps to chatting with AI-powered platforms, children are handing over sensitive data such as names, locations, photos, and even voice recordings without understanding the consequences. This information can be harvested, sold, or used for malicious purposes. Imagine a stranger knowing your child's daily routine, their favorite hangouts, or even their insecurities. It's not just a privacy violation; it's a safety risk. And once the data is out there, it's impossible to take back. As a parent, the thought of my child's personal information being exploited is both infuriating and terrifying.

One of the most dangerous misconceptions parents have is the belief that certain platforms are "safe zones" for children. Let's be clear: no part of the digital world is truly safe. Even kid-friendly apps and games come with risks. Data harvesting, inappropriate content, and predatory behavior can slip through algorithmic filters, exposing your child to harm. The truth is, the internet is not a playground; it's a city without police patrols, where danger lurks around every corner. And your child, trying to navigate it without guidance, is stepping directly into harm's way. As a parent, I've had to confront the uncomfortable truth that even the most innocent-looking apps can pose risks I never anticipated.

The stakes could not be higher. Every second your child spends online is a second they are exposed to potential harm. Cybercrimes can occur in the blink of an eye: a predator sends a message, a scammer tricks them into sharing personal information, a bully posts a humiliating comment. The damage is instant, and the consequences can last a lifetime. The internet is not just a tool; it's a battlefield, and your child is on the front lines.

But here's what I've learned as a parent: we are not powerless. We can't shield our children from every danger, but we can arm them with the tools they need to navigate this digital metropolis safely. We can teach them to recognize red flags, to protect their personal information, and to come to us when something feels off. We can set boundaries, use parental controls, and stay informed about the platforms they use. Most importantly, we can create an environment where they feel safe talking to us about their online experiences, no matter how uncomfortable or scary those conversations might be.

The digital bubble is here to stay, and our children will continue to explore it and grow inside it. But as parents, it's our job to ensure they don't do it alone. We can be their guides, their protectors, and their lifelines in this unpredictable digital world. The risks are real, but so is our ability to fight them. Don't wait until it's too late. Start the conversation today. Your child's safety and their future depend on it.

Part 2: The Road Ahead Family Challenges Through 2055

If today's challenges seem overwhelming, the next three decades will amplify them in ways that are difficult to fully comprehend. Technological evolution is not slowing down; it is accelerating at an unprecedented pace, reshaping every aspect of our lives, including how we connect, learn, and grow as families. Artificial intelligence, VR, BCIs, and increasingly immersive digital ecosystems will redefine family life in ways that are both exciting and deeply concerning. The digital bubble of 2025 will seem simple and outdated compared to the hyper-connected, AI-driven world of 2055. But with these advancements come profound risks that demand our attention now, before they spiral out of control.

Imagine a world where AI doesn't just assist with homework or household tasks but becomes a constant companion for your child as a virtual friend, tutor, and confidant. While this might sound like a parent's dream, it also raises unsettling questions. What happens when children form emotional bonds with AI entities that are designed to manipulate their behavior? What happens when these systems, driven by corporate interests, prioritize engagement over well-being? The line between human and machine interaction will blur, and parents will face the daunting task of ensuring their children develop healthy relationships both online and offline.

VR will further complicate this landscape. VR could become as commonplace as smartphones are today, offering immersive experiences that feel indistinguishable from reality. While this opens up incredible opportunities for education, creativity, and connection, it also poses significant risks. Children could become so immersed in virtual worlds that they struggle to engage with the real one. The concept of "digital addiction" will take on a whole new meaning, as VR environments are designed to be even more captivating and harder to disengage from than today's apps and games. Parents will need to navigate the fine line between allowing their children to explore these new frontiers and ensuring they remain grounded in the physical world.

Meanwhile, the digital ecosystems of the future will be more pervasive and invasive than ever. Smart homes, wearable tech, and biometric data collection will create an environment where every action is tracked, analyzed, and monetized. While this might offer convenience, it also raises serious privacy concerns. Children growing up in this world will have their every move recorded, from their online behavior to their physical location. The concept of privacy as we know it may cease to exist, and parents will need to fight to protect their children's data from being exploited by corporations, governments, or malicious actors.

The social implications of these advancements are equally troubling. As AI and automation reshape the job market, families will face new economic pressures. The skills required to thrive in a digital-first world will evolve rapidly, leaving many parents struggling to prepare their children for an uncertain future. The gap between those who can navigate this new world and those who cannot will widen, magnifying inequality and creating new forms of social division.

And then there's the psychological impact. As children spend more time in digital environments, their ability to form deep, meaningful relationships in the real world may diminish. The constant stimulation of AI-driven content and virtual experiences could erode their capacity for patience, empathy, and critical thinking. Parents will need to work harder than ever to teach values like resilience, curiosity, and emotional intelligence in a world that often prioritizes instant gratification and superficial engagement.

The challenges we'll face over the next three decades aren't just about technology; they're about what it means to be human. They push us to ask tough questions about how families can stay connected and grounded in a world dominated by the digital bubble. The future is coming, ready or not. But if we tackle these challenges head-on, we can make sure technology strengthens family life instead of weakening it. The next three decades will test us, but they also give us a chance to rethink what family means in the digital age. The power to shape that future is in our hands.

Challenge #1: AI-Driven Parenting the Double-Edged Tech Nanny

The next three decades will see AI becoming an integral part of family life, fundamentally reshaping how children are raised and educated. AI companions, digital tutors, and autonomous household assistants will no longer be futuristic concepts; they will be everyday realities, deeply embedded in childhood experiences. While these tools promise convenience, efficiency, and even enhanced learning opportunities, they also raise profound ethical and emotional challenges that families must confront.

Imagine a world where your child's best friend isn't a classmate or neighbor but an AI-powered companion, a virtual entity designed to entertain, educate, and emotionally support them. These AI companions could be incredibly sophisticated, capable of holding conversations, telling stories, and even adapting to a child's personality and needs. The concern is that children may form stronger emotional bonds with these AI entities than with their own families. If a child turns to an AI for comfort, advice, or companionship, the parent-child relationship could become secondary to the algorithms that govern these interactions. The risk is a generation more attached to machines than to the people who love them.

AI-driven education is already transforming how children learn, with personalized lessons, instant feedback, and adaptive curricula. In the future, digital tutors could become so advanced that they replace traditional learning models entirely. While this might seem like a boon for academic achievement, it also diminishes the role of parents in their children's intellectual and emotional development. Parents could become spectators in their children's education, leaving it entirely to AI systems. The danger is that children might excel academically but miss out on the critical life lessons, values, and emotional support that only human guidance can provide.

From smart cribs that rock babies to sleep to AI-powered nannies that monitor and entertain toddlers, autonomous household assistants are poised to take over many aspects of parenting. These tools can make life easier for busy parents, but they also risk creating emotional distance. Digital nannies could become the default caregivers, further reducing the time parents spend bonding with their children. If children grow up relying on machines for comfort and care, they may struggle to form deep, meaningful relationships with their families and peers.

At the heart of these challenges is a fundamental question: Who is really raising our children, parents or algorithms? Without strong parental guidance, families may find themselves raising a generation more influenced by AI systems than by human values. Algorithms are designed to optimize for engagement, efficiency, and convenience, but they lack the moral compass, empathy, and intuition that define human parenting. The risk is a world where children are shaped by data-driven decisions rather than the love, wisdom, and values of their families.

The rise of AI-driven parenting is inevitable, but its impact on families is complex and multifaceted. The challenge lies in navigating a future where technology plays an increasingly central role in childhood, potentially reshaping the very nature of family relationships and human connection.

Challenge #2: Virtual Childhoods Growing Up in the Metaverse

As AR, VR, and MR technologies advance, children and teenagers will increasingly have access to fully immersive digital worlds, spaces where they can play, learn, and socialize in ways that were once the stuff of science fiction. These environments, often referred to as the metaverse, promise to revolutionize entertainment, education, and even socialization. However, they also present profound challenges that could fundamentally alter how they perceive and interact with the world around them.

In the metaverse, the boundaries between the physical and digital worlds will become increasingly blurred. Children and teenagers may find themselves spending hours in hyper-realistic virtual environments, where they can build cities, explore fantastical landscapes, or interact with friends as digital avatars. While these experiences can be enriching, they also risk distorting a child's understanding of reality. Will children struggle to distinguish between virtual achievements and real-life accomplishments? The immersive nature of these worlds could make it difficult for them to separate digital escapism from the tangible experiences that shape personal growth and development.[106]

The metaverse is designed to be captivating, with endless opportunities for exploration, competition, and social interaction. This makes it inherently addictive, especially for young minds that are still developing self-regulation skills. Will children become so immersed in these digital spaces that they neglect their real-world responsibilities and relationships? The risk is a generation that prioritizes virtual achievements, such as leveling up in a game, accumulating digital assets, or gaining social status online, over real-world experiences and connections.

As virtual worlds become more compelling, traditional childhood activities such as playing outside, engaging in creative play, and socializing face-to-face may be forgotten. The physical, sensory, and emotional benefits of these activities are irreplaceable. Running through a park, looking at flowers, or simply laughing with friends in person fosters creativity, resilience, and social skills that cannot be replicated in a digital environment.

Without firm parental intervention, children may grow up more attached to digital realities than to the physical world. The metaverse offers a level of control, customization, and instant satisfaction that the real world simply cannot match. In virtual spaces, children can be whoever they want, achieve whatever they desire, and escape from the challenges and imperfections of real life. The risk is a future where children struggle to find meaning and fulfillment in the real world, having grown accustomed to the limitless possibilities of the metaverse.

The expansion of the metaverse and virtual living represents a profound shift in how children experience the world. While these technologies offer exciting opportunities, they also pose significant risks to the development of healthy, well-rounded individuals.

The challenge lies in navigating this new frontier, ensuring that children can enjoy the benefits of virtual worlds without losing touch with the physical one.

Challenge #3: The Fight for Cognitive Freedom Protecting Young Minds

Within the next three decades, BCIs are poised to revolutionize how we interact with technology, blurring the line between human thought and digital systems. These innovations, which allow direct communication between the brain and external devices, promise groundbreaking possibilities. enhanced learning, seamless communication, and even the ability to control machines with mere thoughts. Yet, they also introduce unprecedented ethical and security risks that could redefine what it means to think, learn, and exist as families.

As BCIs become mainstream, children's cognitive development could be profoundly influenced by external systems. Imagine a world where AI analyzes a child's thoughts to optimize learning or even suggest behavioral adjustments. While this could lead to accelerated intellectual growth, it also raises the specter of a future where children's minds are shaped not by their own experiences and choices, but by algorithms designed to maximize efficiency or conformity. The very essence of individuality, the unique way each person thinks, learns, and grows, could be at risk of being homogenized by external forces.

The integration of BCIs into daily life opens the door to the collection and exploitation of neural data. Companies could use this data to tailor advertisements, manipulate preferences, or even predict behavior with frightening accuracy.[107] Governments might leverage it for surveillance, monitoring thoughts or emotions under the guise of security. The potential for misuse is vast, and the consequences of such intrusions into the most private realm of human existence—the mind—are difficult to overstate. The sanctity of thought, once considered inviolable, could become just another commodity in the digital economy.

As AI augmentation becomes more sophisticated, future generations may increasingly rely on external systems to think, decide, and create. The ability to solve problems, form opinions, and engage in deep, independent thought could diminish as individuals become accustomed to outsourcing cognitive processes to machines. This reliance on AI augmentation risks creating a society where critical thinking, creativity, and emotional intelligence are replaced by algorithmic efficiency. The very qualities that make us human, our capacity for curiosity, imagination, and independent reasoning, could be eroded.

The rise of BCIs presents families with a profound dilemma. On one hand, these technologies offer unparalleled opportunities to enhance human potential. On the other, they threaten to compromise the autonomy and individuality that define our humanity. The digital family of the future will face a critical juncture: to passively accept these shifts, allowing external systems to shape their thoughts and behaviors, or to take an active role in ensuring that technology serves their needs without undermining their autonomy.

The battle for cognitive autonomy is not just a technological challenge; it is a fight for the future of human identity. As we stand on the brink of this new frontier, the choices we make will determine whether we remain the authors of our own thoughts or become mere passengers in a world where the mind is no longer entirely our own. The stakes are nothing less than the preservation of what it means to be human.

Part 3: The Digital Family Playbook

Having explored the challenges families face in 2025 and those anticipated over the next three decades, it's clear that the digital age presents both unprecedented opportunities and profound risks. The future of families will not be determined by technology alone, but by how they choose to engage with it. The key to thriving in this new era is not to reject innovation outright but to harness it wisely, ensuring that technology serves the family rather than controls it.

This section is a practical guide for families navigating the digital bubble. It offers actionable strategies to help you not only survive but thrive in a world where technology is ever-present. From setting boundaries and protecting privacy to fostering real-world connections and educating future generations, these tips empower families to reclaim their autonomy and strengthen their bonds in an era that often seeks to fragment and distract.

The digital world is evolving at breakneck speed, but with the right tools and mindset, families can evolve alongside it, emerging stronger, wiser, and more connected than ever. The journey begins now, and the choices we make today will shape the families of tomorrow. Let's dive into the strategies that will help you take control of your digital life and build a future where technology enhances, rather than undermines, your family's well-being.

Strategy #6: Reclaim Family Autonomy from the Digital Bubble

The first step in reclaiming family life is acknowledging the digital bubble and its pervasive influence. This means recognizing how technology shapes thoughts, behaviors, and family dynamics. For example, social media platforms are designed to keep users scrolling, while gaming apps use psychological triggers to maximize engagement. By understanding these mechanisms, families can begin to resist their pull.[108]

Tip #18

Become a Digitally Conscious Family

Let's face it, digital literacy is no longer optional. It's as essential as teaching your children to tie their shoes or reminding your partner where they left their keys. Every family member, from the toddler who's somehow already better at using your phone than you are to the grandparent who still thinks "Google" is a person, needs to understand how algorithms influence their choices. Because here's the thing: the digital bubble isn't just a fun place to watch videos. It's a carefully engineered system designed to keep you hooked,[109] and if you don't know how it works, you're basically handing over your brain to a bunch of programmers who probably haven't seen sunlight in weeks enjoying office pizza parties.

- **Teach Children How Platforms Use Their Data**

Imagine this: your 10-year-old is scrolling through their favorite app, laughing at memes and watching videos of dogs wearing hats. Meanwhile, in the background, the app is quietly collecting data on what they watch, how long they watch it, and even when they pause to laugh at a particularly silly golden retriever. This data is then used to feed them more of the same content, creating a loop that's hard to break.

Here's a fun way to explain it to your children: "Think of the app like a sneaky little spy. It's watching everything you do so it can figure out how to keep you glued to the screen. And the longer you stay, the more ads it can show you for stuff you don't need, like glow-in-the-dark slime or a $200 gaming chair that you'll use twice and then turn into a laundry holder."

- **Explain Why Certain Content Appears in Their Feeds**

Ever wonder why your kid's feed is suddenly full of videos about unboxing toys or teenagers doing dance challenges? It's not magic—it's algorithms. These digital puppet masters analyze what your child clicks on, how long they watch, and even what they skip to decide what to show them next.

Maybe you can explain it like this: "The app is like a nosy friend who keeps suggesting things based on what you've already said you like. But instead of saying, 'Hey, you liked that video of a hamster eating a sandwich, so here's another one,' it's more like, 'Oh, you watched one hamster video? Here are 47 more, and also some ads for hamster food you don't even need because we don't have a hamster'!"

➢ **Help them Recognize Manipulative Design Features**

Infinite scrolling and autoplay are the digital equivalent of a bottomless bag of chips; you keep going long after you're full because, well, it's just there. These features are designed to keep users engaged by removing natural stopping points.

Here's a funny way to point this out to your children: "You know how you always say, 'Just one more episode,' and then it's midnight and you're crying because the characters in your show just broke up? That's autoplay's fault. It's like a sneaky little gremlin that hits 'next' for you so you don't even have to think about it. And infinite scrolling? That's the gremlin's cousin, who keeps adding more chips to the bag so you never realize you've eaten the whole thing."

➢ **Empower Children to Make Informed Decisions**

Once children understand how the digital bubble works, they can start making smarter choices. For example, instead of mindlessly scrolling, they might decide to set a timer for their screen time or choose to watch content that's actually meaningful to them.

You can frame it like this: "Think of your brain like a castle. Algorithms are like sneaky invaders trying to climb over the walls. But if you know their tricks, you can build a barrier, raise the drawbridge, and say, 'Not today, invaders!' You're the boss of your brain, not some app that just wants you to watch 17 more videos of people falling off skateboards.".

Tip #19
Encourage Transparency and Open Dialogue

Let's be real: trying to police your children's online activities is like herding cats, exhausting, and likely to end with you questioning your life choices. Instead of playing the role of the digital sheriff, for which they will just find ways to outsmart you, why not create a culture of openness where technology is discussed without judgment? This approach not only builds trust but also empowers your children to navigate the digital world with confidence and critical thinking. Plus, it's a lot less exhausting than trying to decode their latest TikTok trend.[110]

- **Encourage Sharing Without Fear**

Children are naturally curious, and the digital world is full of things that spark their interest, whether it's a hilarious meme, a cool DIY project, or a video of a raccoon stealing someone's picnic. But it's also full of stuff that might confuse, upset, or even scare them. The key is to create an environment where they feel safe sharing all of it, the good, the bad, and the weird.

For example, imagine your 12-year-old comes to you and says, "I saw this video where people were eating Tide Pods. Is that… a thing?" Instead of freaking out or, worse, Googling "Tide Pod challenge" and falling down a rabbit hole of parental despair, you could say, "Wow, that sounds bizarre! Let's look it up together and figure out why anyone would think that's a good idea." By responding with curiosity rather than judgment, you're teaching them that it's okay to ask questions and that you're a safe person to turn to when they encounter something strange or troubling online.

- **Turn "Tech Talk" into Family Time**

Make discussing technology a regular part of family life. This doesn't have to be a formal sit-down conversation. Instead, weave it into everyday moments.

For instance, during dinner, you could ask, "What's the funniest thing you saw online today?" or "Did anyone come across something weird or confusing on their phone?" If your teenager rolls their eyes and says, "Ugh, no one talks like that," you can counter with, "Okay, fine. What's the dumbest TikTok trend you've seen this week?" Trust me, there will be no shortage of material.

Another idea: Share your own online experiences. Did you fall for a phishing email? Tell them about it! Did you accidentally like a post from 2017 while stalking an old friend's profile? Admit it! Showing them that you're not perfect and that you can laugh at your own digital missteps makes it easier for them to open up about their own.

➢ **Foster Healthy Skepticism**

The digital world is full of shiny distractions, from viral challenges to "too good to be true" offers. By encouraging your children to question what they see online, you're helping them develop a healthy skepticism that will serve them well in life.

For example, if your child excitedly tells you about a "free" game that requires them to enter their email and password, you can say, "Hmm, that sounds fun, but let's think about why they might want your info. What's in it for them?" This kind of conversation helps them understand that not everything online is as innocent as it seems.

Or, if they're obsessed with a new influencer who's promoting a "miracle" product, you can gently ask, "Do you think they're being honest, or are they just trying to sell something?"

When you create a culture of openness, you're not just teaching your children about technology; you're building a foundation of trust and communication that extends far beyond the digital world. They'll learn that it's okay to make mistakes, ask questions, and seek guidance. And who knows? You might even get a front-row seat to the next big internet trend before it goes viral.

So, the next time your kid starts a sentence with, "So, I saw this thing online…" don't panic. Take a deep breath, put on your best "I'm a cool and approachable parent" face, and dive into the conversation. Because in the digital bubble, curiosity and trust are your family's best defenses.

Tip #20

Set Digital Boundaries for a Balanced Home by Place

Boundaries are the foundation of a healthy relationship with technology. Without them, the digital bubble will overstep on every aspect of family life. Here's how to establish and maintain these boundaries:

➢ **Set Screen-Free Zones**

Let's be honest, trying to have a family dinner while everyone is glued to their screens is frustrating, chaotic, and often feels like an uphill battle that ends with everyone more disconnected than before. That's why designating specific areas of your home as screen-free zones is a game-changer. These spaces become sanctuaries where real-life connections can flourish, free from the constant ping of notifications or the siren call of autoplay.

- **Make Meal Areas a Tech-Free Zone for Real Conversation**

Picture this: you've spent an hour cooking a meal that you're pretty sure qualifies you for a Michelin star. You call everyone to the table, only to find your teenager scrolling through Instagram, your partner answering work emails, and your youngest watching unboxing videos on YouTube. Sound familiar?

By making the dining table a screen-free zone, you're sending a clear message: this is a time to connect, not to scroll. Sure, there might be some initial resistance, but stick to your guns. Over time, you'll find that meals become a time for laughter, storytelling, and maybe even a few heated debates about which singer should have won The Voice last week.[111]

- **Make Your Bedroom a Tech-Free Sanctuary for Rest**

Bedrooms should be a place of rest, relaxation, and maybe a little bit of daydreaming, not a 24/7 command center for managing notifications and late-night scrolling. By keeping screens out of bedrooms, you're creating a space where everyone can unwind and recharge without the blue light glow of doom.

Imagine this: your kid's room is a peaceful haven where they read books, listen to music, or just stare at the ceiling and wonder why socks always disappear in the laundry.[112] And if they ever tell you why, please let me know. Meanwhile, your bedroom is a tranquil retreat where you can finally catch up on sleep instead of falling down a rabbit hole of "just one episode."

Of course, enforcing this rule might require some creativity. For example, you could introduce a "device docking station" in a common area where everyone charges their phones overnight. And if your teenager protests with the classic, "But I need my phone for my alarm!" you can hit them with the ultimate parent power move: "Welcome to the wonderful world of analog clocks of 1995."

Screen-free zones aren't just about reducing distractions; they're about reclaiming your home as a place of connection and calm. They remind your family that there's a world beyond the screen, full of real-life moments worth savoring.

So, go ahead and declare your dining table and bedrooms screen-free zones. Will there be resistance? Probably. Will it be worth it? Absolutely. And who knows? You might even rediscover the joy of a good old-fashioned conversation or at least the satisfaction of knowing you've temporarily outsmarted the digital bubble.

Tip #21

Set Digital Boundaries for a Balanced Home by Time

- **Set Device Curfews**

Let's face it: devices are like that one friend who always overstays their welcome. They're fun at first, but after a while, you're just waiting for them to leave so you can finally get some peace. That's where device curfews come in, a simple but powerful way to reclaim your family's time, attention, and sanity. By setting specific times when all devices are turned off, you're giving everyone's brain a chance to reset, improving sleep quality and creating space for meaningful family moments.

- **Unplug at Night to Sleep Deep and Wake Clear**

We've all been there: you tell yourself you'll just watch one more episode, scroll for five more minutes, or check your email one last time. Next thing you know, it's 2 a.m., and you're deep into a Wikipedia rabbit hole about the history of ketchup.

Device curfews can save you from this fate. Set a rule that all devices are turned off at least an hour before bedtime. This gives everyone's brain a chance to wind down, making it easier to fall asleep and stay asleep. Plus, it creates a buffer zone where you can read a book, chat with your partner, or just stare at the ceiling and wonder why you thought it was a good idea to binge-watch that entire series in one night.

For children, this is especially important. The blue light from screens can mess with their sleep hormones, making it harder for them to fall asleep and stay asleep. By enforcing a device curfew, you're not just improving their sleep quality; you're also giving them a chance to develop healthier bedtime habits.[113]

Device curfews aren't just about reducing screen time; they're about creating boundaries that protect your family's well-being. They give everyone a chance to disconnect from the digital world and reconnect with each other and themselves.

- **Encourage Digital Fasting**

Introducing regular periods of digital fasting, where the entire family disconnects from screens for a set amount of time, can be a game-changer for mental well-being and family dynamics. These breaks, whether they last a few hours, a full day, or even an entire weekend, provide a much-needed reset for the brain.[114] Without the constant buzz of notifications or the pull of endless scrolling, everyone gets a chance to recharge, refocus, and reconnect with what truly matters.

For example, you could designate Sunday afternoons as a screen-free time, where the family engages in activities like board games, cooking together, or going for a walk. Or, you might try a "digital detox weekend" where phones, tablets, and TVs are off-limits, and instead, you focus on outdoor adventures, creative projects, or simply enjoying each other's company.

These digital fasts not only help improve focus and reduce stress but also create space for meaningful conversations and shared experiences. Over time, they can become a cherished family tradition, a reminder that life is richer and more fulfilling when we step away from the screens and truly engage with the world and each other. Plus, it's a great opportunity to show your children and yourself that fun and connection don't require a Wi-Fi connection.

Tip #22

Practice mindful consumption

Teach children to recognize when they're being drawn into an algorithmic loop of endless scrolling or autoplay. Encourage them to ask themselves, Am I using this device intentionally, or is it using me? This awareness helps them develop healthier habits and resist the pull of mindless consumption.

The digital world is like a never-ending buffet of distractions. One minute, you're watching a video about how to fold a fitted sheet, and the next, you're three hours deep into conspiracy theories about why pigeons are government spies. For children, it's even worse; algorithms are designed to keep them hooked with endless loops of autoplay, infinite scrolling, and content that's just one more click away.

That's where mindful consumption comes in. It's about teaching children to recognize when they're being sucked into the digital vortex and helping them take back control. The goal? To shift from mindless scrolling to intentional use so they're the ones in charge, not the algorithm.

➢ **Teach Children to recognize the signs of algorithmic Loops**

Algorithms are sneaky. They're like that friend who keeps saying, "Just one more episode!" even though you both know you have work in the morning. For children, this can look like endless scrolling through TikTok, binge-watching YouTube videos, or getting lost in a game that never seems to end.

The first step to breaking free is awareness. Teach your children to recognize the signs that they're in an algorithmic loop. Are they scrolling just because the next video automatically plays? Are they watching something they don't even care about, just because it's there? Help them pause and ask themselves, Am I using this device intentionally, or is it using me?

➢ **Help Children master the art of making a Pause**

One of the best tools for mindful consumption is the simple act of pausing. Encourage your children to take a breath before they click, scroll, or swipe. Ask them: Do I really want to watch this, or am I just bored? Is this adding value to my day, or is it just filling time?

For example, if your child is about to click on yet another "recommended" video, you can gently remind them, "Hey, is this something you actually want to watch, or is the app just trying to keep you hooked?"

➢ **Build Healthier Digital Habits**

Mindful consumption isn't about cutting out screens entirely; it's about using them with purpose. Encourage your children to set goals for their screen time. Maybe they want to watch a tutorial to learn a new skill, or maybe they're using an app to connect with friends. The key is to be intentional.

You can even turn it into a family challenge. For example, see who can go the longest without falling into an autoplay trap. Teaching children to consume content mindfully isn't just about reducing screen time; it's about empowering them to make smarter choices. It's about helping them develop the critical thinking skills they need to navigate a world that's constantly vying for their attention.

Tip #23
Resist the Pull of Algorithmic Manipulation

Algorithms are designed to keep users engaged at all costs, often at the expense of their well-being. These systems, powered by artificial intelligence, are finely tuned to capture attention, exploit psychological vulnerabilities, and maximize screen time, whether through endless scrolling, personalized content feeds, or addictive notifications. For families, this can lead to fragmented attention, reduced quality time, and even mental health challenges, as the digital bubble increasingly encroaches on daily life.

However, families can fight back by taking control of their digital environments. By adjusting configurations and tweaking settings, they can reclaim their time, focus, and well-being. Here are some actionable steps:

- **Disable Autoplay and Push Notifications**

Disabling autoplay is a powerful yet simple step families can take to reduce the grip of technology on their attention and time. These features, while seemingly convenient, are deliberately designed to keep users hooked, often leading to mindless consumption and distraction. By turning them off, families can create a more intentional and controlled digital environment.

Autoplay is a common feature on streaming platforms like *YouTube*, *Netflix*, and social media apps. It automatically plays the next video or episode, making it easy to lose track of time and fall into endless scrolling or binge-watching. Here's how to tackle it:

For example, on Netflix, go to your account settings and disable autoplay for the next episode in a series. On YouTube, turn off autoplay by clicking the autoplay toggle. This small change encourages more mindful viewing and helps family members decide when to stop rather than being pulled into an endless loop of content.

As for platforms like *Facebook* and *Instagram* also use autoplay for videos in feeds. Disable this feature in the app settings to prevent videos from playing automatically as you scroll. This reduces distractions and helps users focus on the content they intentionally choose to engage with.

➢ Disable Push Notifications

Push notifications are also one of the most effective tools apps use to pull users back in. Whether it's a news alert, a social media update, or a game reminder, these notifications are designed to interrupt your day and demand attention. Here's how to take control:

Go through your phone's notification settings and disable push notifications for apps that don't require immediate attention. For example, turn off notifications for social media platforms, games, or shopping apps. This reduces the constant buzzing and pinging that can disrupt family time, work, or relaxation.

Keep notifications enabled only for essential apps, such as messaging platforms (e.g., *WhatsApp* or SMS) or calendar reminders. This ensures you stay connected to what truly matters without being overwhelmed by unnecessary interruptions.

Some devices allow you to schedule "Do Not Disturb" modes or batch notifications so they arrive at specific times. For example, you can set your phone to only deliver notifications during certain hours, allowing for uninterrupted family dinners or focused work sessions.

➢ Follow Human-Curated Content

Instead of relying solely on AI-driven recommendations, which are often designed to maximize engagement rather than provide meaningful content, families can seek out human-curated resources that align with their values and interests. This shift not only fosters a healthier media diet but also encourages critical thinking and intentional consumption. By prioritizing quality over quantity, families can break free from the algorithmic echo chambers that often prioritize sensationalism and engagement over substance.

➢ Use Platforms That Prioritize Quality Over Quantity

When it comes to entertainment, choosing platforms that prioritize quality and curation can make all the difference. Look for services that offer thoughtfully selected films, documentaries, and educational content, often available through partnerships with libraries or dedicated streaming platforms. Some platforms specialize in documentaries and educational programming, covering fascinating topics like science, history, and technology. For families with children, there are resources that provide reviews and recommendations for age-appropriate movies, TV shows, books, and apps, helping parents make informed decisions that align with their family's values. By opting for these curated options, you can ensure that your entertainment choices are not only enjoyable but also meaningful and enriching for the whole family.

Several platforms are designed to deliver high-quality, human-curated content, offering a refreshing alternative to the overwhelming flood of endless options found elsewhere. These platforms focus on depth and intentionality, making it easier to find meaningful material that aligns with your interests. For example, some websites specialize in hosting in-depth articles written by experts and independent writers, covering a wide range of topics from science to culture. Others offer curated newsletters on niche subjects, crafted by journalists and thought leaders, delivering thoughtful insights straight to your inbox. Additionally, there are tools that help you save and organize content—like articles, videos, and podcasts—allowing you to focus on intentional consumption rather than getting lost in mindless scrolling. By exploring these platforms, you can transform your media habits and create a more purposeful and enriching digital experience.

- **Subscribe to Trusted Newsletters**

One of the easiest and most effective ways to access high-quality, human-curated content is by subscribing to trusted newsletters. These gems of the digital world deliver thoughtful, well-researched content straight to your inbox, perfectly aligned with your family's values and interests. The best part? It's a one-time effort! Once you've chosen a few newsletters that resonate with you, the job is done; no more endless searching or sifting through low-quality content. Sit back, relax, and let the good stuff come to you, making it effortless to stay informed and inspired. It's a small step with a big payoff for your family's media habits!

- **Follow Educators and Thought Leaders on Social Media**

Social media can be a powerful and enriching tool when used with intention. Instead of getting lost in endless scrolling, focus on following educators, scientists, and thought leaders who share valuable insights and resources. Look for accounts that specialize in topics like science, history, technology, or personal development; these often provide educational and inspiring content that can spark curiosity and learning. On platforms like YouTube, seek out channels dedicated to educational videos that cover a wide range of subjects, from science to the arts. On Instagram and Twitter, follow experts in fields that interest your family, whether it's space exploration, environmental science, or philosophy. These accounts offer a steady stream of high-quality, thought-provoking material that can transform your social media feed into a source of knowledge and inspiration. By curating who you follow, you can turn social media into a tool for growth and discovery.

- **Engage with Local and Community Resources**

Local resources can also provide valuable, human-curated content. Many libraries offer curated book lists, workshops, and events tailored to different age groups and interests. Joining or starting a family book club can encourage shared reading and discussion of books that align with your values. Additionally, local museums and cultural institutions often share updates on exhibits, virtual tours, and educational programs through newsletters or social media, offering enriching experiences for the whole family.

- **Create a Family Media Plan**

To make the most of human-curated content, involve the whole family in creating a media plan. Curate a shared playlist or watchlist of movies, shows, or podcasts that everyone can enjoy and learn from. Set aside time to discuss the content you consume, reflecting on what you've learned and how it aligns with your family's values. This not only enriches your media consumption but also fosters meaningful conversations and shared learning experiences.

By seeking out human-curated content, families can break free from the algorithmic traps that often prioritize sensationalism and engagement over quality. This approach ensures that technology serves as a tool for growth, connection, and enrichment, rather than a source of distraction or manipulation. It's a powerful way to cultivate a healthier, more intentional relationship with media and technology.

Tip #24
Turn Off Unnecessary Data Tracking

Reviewing and adjusting privacy settings on family devices is a vital step in protecting personal information and reducing the risks of manipulation and exploitation. While this task may seem overwhelming, seeking help from a trusted tech-savvy individual or a professional can make it manageable. Many apps and platforms collect excessive data such as location, browsing history, and personal preferences, often without users' full awareness. This data fuels targeted advertising, algorithmic content feeds, and other manipulative practices designed to keep users engaged. By taking control of privacy settings, families can limit data tracking and create a safer, more secure digital environment.

Data tracking powers algorithms that manipulate user behavior, from personalized ads to addictive content feeds. By reducing data collection, families can minimize exposure to these exploitative practices, fostering a more intentional and less intrusive online experience. This empowers individuals to take charge of their digital interactions rather than being passively influenced by algorithms.

Moreover, limiting data collection enhances security. Less data stored means fewer opportunities for breaches or misuse, protecting families from identity theft, scams, and other cyber threats. In an era of frequent data breaches, proactively restricting data tracking is a powerful way to safeguard personal and financial information.

By adjusting privacy settings, families can build a safer digital space that prioritizes privacy, reduces manipulation, and strengthens security. This proactive approach ensures technology serves their needs rather than exploiting their data, helping them cultivate a healthier, more intentional relationship with the digital world.

To enhance digital privacy and security, families must take proactive steps to review and adjust settings across various devices and platforms. Implementing these changes can significantly reduce data exposure and minimize digital vulnerabilities. Here's a practical guide to refining privacy settings across smartphones, social media, web browsers, smart home devices, and gaming consoles.

➢ **Smartphones and Tablets**

One of the biggest sources of data collection, smartphones and tablets constantly gather information through location tracking, app permissions, and advertising IDs. Start by reviewing which apps have access to location data in your device settings. While a weather app may need location access to provide accurate forecasts, social media apps and mobile games typically do not. Disable location tracking for unnecessary apps to limit excessive data collection.

Next, audit app permissions, particularly access to sensitive features like the camera, microphone, and contacts. For example, a photo-editing app may require camera access, but a simple puzzle game should not. Revoke permissions for apps that don't need them, reducing the risk of personal data exposure. Additionally, both Android and iOS allow users to reset or disable their advertising ID, which prevents advertisers from tracking activity across apps. By doing so, families can restrict the amount of behavioral data collected and used for targeted advertising.

➢ **Social Media Platforms**

Social media platforms collect vast amounts of personal information, often sharing it with third parties. On *Facebook*, navigating to the "Settings & Privacy" section and using the "Privacy Checkup" tool can help disable data sharing with third-party apps, limit who can see posts, and turn off facial recognition.

On *Instagram*, visiting "Settings" then "Privacy" allows users to adjust visibility settings, control tagging permissions, and restrict messaging from unknown contacts. Disabling the "Activity Status" feature can also prevent others from tracking when a user is online.

For *TikTok*, privacy adjustments are crucial, especially for younger users. In the app's settings, disabling "Personalized Ads" and limiting data sharing with third parties can help minimize tracking. Making accounts private ensures better control over who can view content and interact with family members online.

- **Web Browsers**

Online tracking extends beyond social media and into everyday web browsing. To enhance privacy, enable "Do Not Track" features in browser settings and block third-party cookies. For example, in *Google Chrome*, users can navigate to "Settings" then "Privacy and Security" then "Cookies and Other Site Data" to manage these preferences.

Switching to privacy-focused search engines such as *DuckDuckGo* can also provide an extra layer of security by ensuring that searches remain untracked and personal data isn't stored. These small adjustments can drastically reduce the amount of data being shared with advertisers and third-party trackers.

- **Smart Home Devices**

Smart home devices such as voice assistants and smart TVs often collect more data than most users realize. For devices like *Amazon Alexa* or *Google Home*, reviewing privacy settings and disabling voice recording storage can prevent sensitive conversations from being stored and analyzed. Additionally, regularly deleting existing recordings further enhances privacy.

Smart TVs are another overlooked source of data collection. Many manufacturers track viewing habits to serve personalized advertisements. Navigating to the TV's settings menu and disabling ad tracking can help prevent unnecessary data collection and ensure that household viewing behaviors remain private.

- **Gaming Consoles**

With online multiplayer gaming and social networking features integrated into gaming consoles, privacy concerns extend to devices like *Xbox* and *PlayStation*. Reviewing privacy settings to limit data collection, disabling personalized ads, and restricting communication settings can help protect family members, especially younger users, from exposure to online risks.

By taking these steps, families can create a more secure digital environment, reducing the risk of data exploitation and ensuring greater control over their online presence. Small changes in privacy settings can make a significant difference in safeguarding personal information and maintaining digital well-being.

Tip #25

Build Mental Resilience Against Digital Overload

The constant stimulation of the digital bubble can take a toll on mental health, especially for children. Building resilience is key to maintaining balance and well-being:

In the digital age, emotional intelligence is a critical skill that helps children navigate the complexities of social media and online interactions. Teaching them how to manage their emotions, develop self-awareness, and resist the trap of digital comparison equips them with the tools to build confidence and self-worth from within rather than seeking external validation. Here's how parents and caretakers can foster emotional intelligence in their children:

- **Encourage Open Discussions About Social Media's Influence**

Children should understand that social media is designed to keep them engaged by triggering emotional responses, often through curated content that fuels comparison. You can create a habit of discussing how these platforms work, explaining that likes, comments, and shares are not true measures of success or self-worth. For instance, a family conversation could highlight how a post's popularity does not determine its value, reinforcing the idea that self-esteem should be built on real-world experiences rather than digital approval.

- **Help Children Recognize and Process Their Emotions**

Developing emotional intelligence starts with identifying and understanding emotions. When children feel frustrated, excluded, or anxious after spending time online, you should encourage them to express these feelings and reflect on their triggers. Instead of dismissing their concerns, validate their emotions by saying, "I understand that seeing that post made you feel left out. Let's talk about why and how we can handle those feelings." This teaches children to manage their emotions rather than allowing social media to dictate their mood.

- **Shift the Focus to Personal Growth and Achievements**

A child's confidence should stem from their efforts and accomplishments rather than digital validation. You can reinforce this by making a habit of celebrating personal milestones, big or small. A great way to do this is by introducing a daily or weekly family tradition where each member shares a personal success, such as improving in a sport, learning a new skill, or helping a friend. These conversations help children recognize their own progress and value their journey rather than comparing themselves to others online.

- **Teach Critical Thinking About Online Content**

Social media rarely reflects reality. You can teach children to view online content with an insightful eye by showing them how influencers carefully craft their images and videos. Watching a "behind-the-scenes" clip of a perfectly staged photo or discussing how filters distort reality can help children realize that what they see online is often exaggerated or entirely manufactured. This understanding helps them resist the pressure to meet unrealistic digital standards.

Tip #26

Model Healthy Digital Habits

Alright, confession time—I'm definitely guilty of this one. I, too, have fallen into the spiral of mindless scrolling, checking notifications like my life depends on it, and feeling an irrational sense of victory when a post gets more likes than expected. And guess what? Children are watching everything. They're like tiny, highly observant sponges, soaking up every digital habit we model.

So, I'm doing my best to change that. Instead of being glued to my phone, I'm making an effort to show that real life is where the good stuff happens, putting devices down during family time, actually listening without half my brain in cyberspace, and celebrating accomplishments for what they are, not for how many thumbs-up they get online. Even small things, like casually mentioning how satisfying it was to finish a project without needing the internet's validation, help reinforce that success isn't measured in likes.

Progress is slow, and I won't pretend I don't occasionally slip back into my notification-checking reflex, but if the children are learning by example, I'd rather they copy my efforts to improve than my habit of "just one more scroll" which, let's be honest, is never just one.

- **Promote Gratitude and Self-Reflection**

A strong sense of self-awareness can counteract the negative effects of social comparison. Encouraging children to keep a gratitude journal where they list things they appreciate about themselves and their lives can shift their focus from external validation to personal fulfillment. Reflecting on what truly makes them happy, beyond the digital world, helps reinforce a mindset of self-acceptance and appreciation for life's small but meaningful moments.

By prioritizing emotional intelligence, you can help your children develop resilience against the pressures of the digital landscape. These skills not only safeguard their mental health but also empower them to navigate social media with confidence, ensuring they remain in control of their self-worth rather than being dictated by online validation.

- **Promote Offline Hobbies and Outdoor Activities**

Counterbalancing screen time with activities that engage the mind and body in the real world is essential for maintaining a healthy, balanced lifestyle, especially in today's digital-heavy environment. While screens can be entertaining and even educational, they often pull us into a passive, sedentary mode. To break free from the digital bubble, it's important to actively seek out experiences that stimulate creativity, build resilience, and bring genuine joy. Here's how you can make it happen:

- **Get Moving—Because the Couch is Not Your Friend**

Instead of sinking into the sofa for another marathon of "just one more episode," get the whole family involved in a sport or physical activity. Whether it's football, rock climbing, or ninja warrior training in your backyard, moving your body boosts energy, improves focus, and helps everyone sleep better. Plus, chasing your children around the park totally counts as cardio.

- **Unleash Your Inner Picasso**

You don't have to be the next Van Gogh to enjoy painting, sculpting, or crafting. Grab some supplies and let your family's creativity run wild. Doodle, build, or even try a Pinterest-inspired DIY project. The goal isn't perfection, but it is laughter, self-expression, and proving to yourself that glue sticks can, in fact, end up everywhere.

- **Explore the Great Outdoors**

Take a family hike, explore a local park, or just lie on the grass and name weirdly shaped clouds. Nature has a way of resetting our overstimulated brains.

- **Engage in Hands-On Learning**

Cooking with kids can be a hilarious disaster, but it's also an amazing way to bond. Try new recipes, let them mix and spill ingredients, and celebrate small victories, like successfully flipping an omelet without breaking it. Teaching children to cook not only gives them life skills but also keeps them away from screens while filling the house with the smell of something delicious.

- **Play Board Games**

Dust off the board games and rediscover the joy of good old-fashioned competition. Whether it's the strategic intensity of chess or Scrabble or the absolute chaos of Uno, family game nights create laughter, critical thinking, and some very passionate rule interpretations.

- **Encourage Reading**

Reading isn't just about school assignments; it's a passport to imagination. Whether it's comic books, novels, or goofy bedtime stories, reading fuels creativity while giving eyes a much-needed break from screens.

Balancing screen time isn't about banning technology; it's about making space for real-world joy. The best memories aren't made in front of a screen; they're made in shared laughter, spontaneous adventures, and messy, unpredictable moments. So, let's stop being slaves to the scroll and start living in the moment. After all, when was the last time binge-watching cat videos felt as good as dancing in the kitchen or hiking up a mountain? Exactly.

Strategy #7: Stay Secure—Navigate the Digital Bubble Without Falling into Its Traps

The digital world is not a harmless playground; it's a vast, shadowy landscape filled with hidden threats at every turn. Every click, every message, and every online interaction carries risks: cybercriminals waiting to strike, AI-powered scams designed to deceive, and manipulative tactics that prey on trust and curiosity. It's a place where danger often hides in plain sight, masked by the convenience and allure of technology.

The internet is not a safe, controlled environment; it's unpredictable, unregulated, and, at times, downright dangerous. As parents, it's our responsibility to act as digital guardians, ensuring our children don't wander this risky terrain unprotected.

This isn't just about setting screen time limits; it's about making every moment online as safe as possible. It's about teaching children to recognize threats, avoid pitfalls, and navigate the digital world with confidence and caution. Because in a space where a single click can lead to harm, caution isn't just important, it's essential. The stakes are too high to leave anything to chance. Now let's explore who we stand against those dangers.

Tip #27

Active Parental Involvement Over Passive Surveillance

Think of the internet like a bustling city; some areas are safe and family-friendly, while others are filled with hidden dangers. Just as you wouldn't allow your child to explore an unfamiliar city alone, they shouldn't be left to navigate the digital space without guidance.

While parental control apps and monitoring tools can provide a helpful layer of protection, they're not a replacement for open, honest conversations with your children. Children need to understand why certain precautions are necessary, not just follow rules because a piece of software says so. Think of it like teaching them to cross the road safely; you don't just tell them to look both ways; you explain why it's important to avoid danger.

Here are some practical tips to make digital safety discussions a natural part of your family routine:

➢ **Explain the "Why" Behind Rules**

Instead of just saying, "Don't talk to strangers online," explain how predators or scammers might try to trick them. Use age-appropriate examples to help them understand the risks.

➢ **Role-Play Scenarios**

Practice what to do if they encounter something suspicious online, like a stranger asking for personal information or a pop-up offering a "free prize." This helps them feel prepared and confident.

- **Talk About Privacy**

Teach them why it's important to keep personal information private, like their full name, address, or school. Explain how sharing too much can put them and the family at risk.

- **Discuss Cyberbullying**

Talk about what cyberbullying looks like and how to respond if they or a friend experience it. Emphasize the importance of kindness and standing up for others.

Tip #28

Limit Exposure to Online Strangers and Set the Boundaries Together

Involve your children in setting rules for screen time, app usage, and online behavior. When they feel part of the process, they're more likely to follow the guidelines.

By having these ongoing conversations, you're not just protecting your children; you're empowering them to make smart, safe choices online. After all, the best tool for navigating the digital world isn't an app; it's a well-informed, confident child who knows how to stay safe.

The internet has a unique way of blurring the lines between friend and foe. On gaming platforms, in chat rooms, and across social media sites, children can interact with people from all over the world, many of whom may not be who they claim to be. A friendly gamer offering tips could be a predator in disguise, or a seemingly harmless chat could lead to dangerous situations. This is why it's crucial to teach children the golden rule of online interactions: If you wouldn't talk to them in real life, don't engage online. Here's how to help children understand and apply this rule:

- **Explain the Risks of Anonymous Interactions:**

Talk to your child about how people online can hide behind fake names, photos, and profiles. Explain that just because someone seems nice or shares similar interests doesn't mean they're trustworthy. Use relatable examples, like how they wouldn't strike up a conversation with a stranger in a park without knowing who they are.

- **Teach Children to Recognize Red Flags**

Help them identify warning signs, such as someone asking for personal information, pressuring them to keep a conversation secret, or trying to move the interaction to a private platform. Role-play scenarios to practice how they should respond.

- **Set Clear Boundaries for Online Friendships**

Encourage them to only interact with people they know in real life or who have been vetted by you. If they meet someone new online, they should always check with you before continuing the conversation.

- **Use Gaming and Social Media as Teaching Tools**

If your child plays online games or uses social media, review their friend lists or connections together. Ask questions like, "Do you know this person in real life?" or "How did you meet them online?" This helps them think critically about their connections.

- **Reinforce the "Golden Rule" Regularly**

Make it a habit to remind your child of the golden rule before they go online. For instance, say, "Remember, if you wouldn't talk to someone at school or in the neighborhood, don't talk to them online either."

- **Discuss the Consequences of Engaging with Strangers**

Use age-appropriate examples to explain how engaging with strangers online can lead to scams, bullying, or even dangerous situations. Help them understand that their safety is more important than being polite or curious.

By teaching children to apply the golden rule of online interactions, you're giving them a simple yet powerful tool to navigate the digital world safely. It's not about instilling fear but about empowering them to make smart, confident choices. After all, the internet may blur the lines between friend and foe, but with the right guidance, your child can learn to see clearly and stay safe.

Tip #29
Prevent Cyber Attacks and Identity Theft

Cybercriminals don't just target large corporations; they target children too. Children and teenagers often unknowingly expose personal data through social media posts, gaming accounts, and unsecured websites. Bad actors use this data for identity theft, scams, or even worse.

- **Fortify Family Accounts**

Fortifying family accounts is a crucial step in safeguarding your digital life from hackers, scams, and unauthorized access. Start by using strong, unique passwords for every account, and never reuse them across platforms. Reusing passwords is like using the same key for your house, car, and office; if someone gets hold of it, everything is at risk. Create passwords that combine letters, numbers, and symbols, and consider using a password manager to securely store and generate them. For example, instead of something simple, opt for a more complex password like. This small change can make a big difference in keeping your accounts secure.

Another essential layer of protection is enabling Two-Factor Authentication (2FA) on all devices and accounts. Even if someone guesses or steals your password, 2FA adds an extra step by requiring a second form of verification, such as a code sent to your phone or generated by an authentication app. This simple step can prevent unauthorized access and give you peace of mind. Apps like *Google Authenticator* or *Authy* make it easy to set up and manage 2FA across multiple accounts. For instance, when logging into an account, you'll enter your password and then a unique code sent to your device, ensuring that only you can access your information.

By taking these steps, you're building a digital fortress that protects your family's online presence. It's a small investment of time that pays off in security and peace of mind while also setting a great example for your children about the importance of staying safe in the digital world.

- **Teach Children to Be Data-Conscious**

Teaching children to be data-conscious is an essential part of helping them navigate the digital world safely. Many children don't realize how much personal information they unintentionally reveal online, whether through gamertags, usernames, or social media bios. These details, like real names, birthdays, or even the name of their school, can be pieced together by strangers to create a surprisingly clear picture of their lives. To protect their privacy, encourage them to use anonymous usernames in games and on social platforms.

It's also crucial to reinforce the rule that they should never share personal details with online friends, no matter how friendly or trustworthy they seem. Explain that people online aren't always who they claim to be, and even seemingly harmless information such as their favorite hangout spot or the name of their pet can be used in harmful ways. Role-play scenarios to help them practice responding to questions like, "What school do you go to?" or "Where do you live?" with answers like, "I don't share that online."

By teaching children to be mindful of what they share and how they present themselves online, you're equipping them with the tools to protect their privacy and stay safe. This awareness not only helps them avoid potential dangers but also fosters a sense of responsibility and confidence as they navigate the digital world.

➢ **Recognize and Avoid Phishing Scams**

Recognizing and avoiding phishing scams is a critical skill for staying safe online, especially for children who may not yet be familiar with the tactics cybercriminals use. Phishing scams often involve cybercriminals disguising themselves as legitimate businesses, schools, or even friends to trick people into revealing personal information or downloading malicious software. Teach your children to be cautious and skeptical of anything that seems out of the ordinary. For example, if they receive an email claiming they've won a prize or a message from a "friend" asking for sensitive information, they should pause and think before clicking.

Encourage them to never click on suspicious links, open unknown emails, or download random attachments, even if they appear to come from someone they know. Explain that cybercriminals can fake email addresses and create messages that look real but are designed to steal information or harm their devices. A good rule of thumb to share with them is, if it sounds too good to be true, it's a scam. Whether it's a message promising free gaming currency, a notification about a fake contest win, or an urgent request for personal details, they should always double-check with a trusted adult before taking any action.

By teaching children to recognize the signs of phishing scams and respond with caution, you're helping them build a strong defense against one of the most common online threats. This awareness not only protects their personal information but also empowers them to navigate the digital world with confidence and critical thinking.

Tip #30
Handle Cyberbullying and Online Conflicts

The digital world has amplified the impact of bullying. Unlike in-person conflicts that end when children leave school, cyberbullying follows them 24/7, invading their homes through screens. Knowing how to respond and resist is crucial.

➢ Silence and Blocking Are Powerful Weapons

Dealing with online bullies can be tricky, but one of the most effective strategies is to avoid engaging with them altogether. Responding to bullies, whether by arguing or defending yourself, often fuels their behavior and gives them exactly what they want: a reaction. Instead, teach children to use the block button and ignore provocations. Online aggressors thrive on attention, and by refusing to give them that satisfaction, your child takes away their power.

Encourage children to see the block button as a tool for taking control of their online experience. If someone is being mean, disrespectful, or overly aggressive, they don't have to tolerate it; they can block the person immediately. This not only stops the bully from contacting them but also sends a clear message that their behavior won't be tolerated. Additionally, teach them to avoid responding to hurtful comments or messages, even if it's tempting to defend themselves. Explain that bullies often escalate their behavior when they get a reaction, so silence can be the most powerful response.

It's also important to remind children that they don't have to handle these situations alone. Encourage them to come to you or another trusted adult if they're being targeted online. By combining the power of the block button with the support of caring adults, children can effectively shut down bullies and protect their mental well-being. This approach not only helps them navigate online conflicts but also builds resilience and confidence in handling difficult situations.

➢ Encourage Reporting

Many platforms today come equipped with built-in reporting features specifically designed to address cyberbullying, making it easier for users to flag harmful behavior and seek help. Whether it's a social media site, a gaming platform, or an educational app, these tools are there to protect users and hold bullies accountable. Schools, social media platforms, and even law enforcement can intervene in more serious cases, ensuring that the issue is addressed appropriately. It's important to make sure your child understands that reporting cyberbullying isn't snitching or causing unnecessary trouble; it's an act of self-protection and a way to stand up for themselves and others.

Explain to your child that reporting harmful behavior helps create a safer online environment for everyone. Just as they would tell a teacher or parent if someone was being mean to them in person, reporting online bullying is no different.[115] Encourage them to use these features confidently, knowing that they're taking a responsible and courageous step. For example, if they receive a hurtful message or see someone being targeted in a group chat, they can use the platform's reporting tools to alert moderators or administrators.

By empowering your child to use these tools and seek help when necessary, you're teaching them to advocate for themselves and others, fostering a sense of responsibility and resilience in the face of online challenges.

➢ **Promote Digital Empathy**

Promoting digital empathy is crucial in helping children understand that there's a real person behind every screen, even if they can't see them. The anonymity of the online world can make it easy to forget that words have power and can hurt just as much, if not more, than they do in real life. Teach children to pause and think before they type, encouraging them to consider how their words might affect others. For example, ask them how they would feel if someone sent them a mean comment or excluded them from an online group. This helps them connect their actions to real emotions and consequences, fostering a sense of responsibility and kindness in their digital interactions.

Encourage them to practice respect and empathy in all their online activities, whether they're chatting with friends, playing games, or commenting on social media. Remind them that once something is posted or sent, it can't be taken back, and it can spread quickly, causing lasting harm.[116] By fostering a mindset of digital empathy, you're helping them become thoughtful, compassionate digital citizens who contribute to a kinder and more positive online environment

Tip #31
Recognize the Threat of AI-Powered Manipulation

AI isn't just here to make life easier; it's also a master of influence, and not always in ways that work in your or your family's favor. Think of it like that "helpful" friend who insists they know what's best for you but secretly has their own agenda. For children and teenagers, who are still figuring out how to navigate the world, AI can be especially sneaky.[117] From deepfake videos that make it hard to tell what's real to algorithms that know exactly how to keep them scrolling (or shopping), AI has a knack for pushing buttons they didn't even know they had. And let's not forget the emotionally manipulative content meant to make people upset or overly emotional, like a 13-year-old getting angry over a fake news story.

The scary part? Children often don't realize they're being manipulated. They might think they're just watching a funny video or playing a game, but behind the scenes, AI is busy learning how to keep them hooked. It's like a digital puppet master, pulling strings to shape their opinions, habits, and even their self-esteem. The good news? By teaching children to question what they see online and understand how AI works, we can help them spot the strings before they get tangled. After all, the best defense against a sneaky AI is a curious, critical mind and maybe a healthy dose of skepticism. Because if something online seems too perfect, too shocking, or too good to be true, chances are, it's not just magic; it's algorithms.

- **Verify for Deepfakes & AI-Generated Misinformation**

AI has reached a level of sophistication where it can create fake videos, images, and news articles that look and sound shockingly real. Deepfakes and AI-generated misinformation are like the ultimate magic trick, except the magician is a computer, and the audience is everyone on the internet. For children, who are already navigating the wild world of social media and online content, this can be especially confusing. Imagine trying to figure out if that video of a celebrity saying something outrageous is real or just a clever digital puppet show.

Teaching children to question and verify online content is more important than ever. Encourage them to think critically about what they see and hear online. If something seems too shocking, too convenient, or just too good to be true, it probably is. A good rule of thumb is to fact-check before sharing or believing. For example, if they come across a video of their favorite YouTuber doing something completely out of character, they should consider whether it makes sense, check trusted news sites, and assess if there's a chance it could be fake.

Fact-checking isn't always easy, especially when the fake stuff looks so convincing. But that's where a healthy dose of skepticism comes in handy.[118] Teach your children to treat online content like a mystery they need to solve. They can ask who created it, why, and what the source is. It's like being a digital detective, except instead of a magnifying glass, they've got *DuckDuckGo* and a critical mind. And if they're ever unsure, remind them it's okay to ask for help. After all, even Sherlock Holmes had Watson.

By equipping children with the tools to question and verify, you're not just helping them avoid falling for fake news or deepfakes. You're teaching them to be smarter, savvier digital citizens. And who knows? Maybe they'll even start fact-checking your stories at the dinner table.

- **Be aware of Emotional AI Manipulation in social media & Gaming**

Social media platforms and gaming apps have become masters of emotional manipulation, thanks to AI-driven dopamine triggers like likes, streaks, and loot boxes. These systems are carefully designed to keep users hooked by tapping into their emotions, making children feel compelled to stay engaged for hours. It's like a digital carnival where every click, swipe, or tap gives you a tiny reward, but the price is your family time, attention, and sometimes even their peace of mind. For children, who are still developing self-control, these tactics can be especially hard to resist. And here's the kicker: this trend is only expected to grow bigger over the next three decades. As AI becomes even more sophisticated, the emotional hooks will only get harder to spot and even harder to resist.

Teaching children to recognize these manipulative tactics is crucial. Help them understand that those little red notification bubbles or the promise of a rare in-game item aren't just fun features; they're carefully crafted to keep them coming back for more. Explain how these systems work, using relatable examples like how a slot machine keeps players pulling the lever or how a candy bar is designed to make you crave another bite. Once they see the mechanics behind the magic, they'll be better equipped to resist the pull.

Encourage them to ask themselves, "Am I having fun, or am I just chasing the next reward?" By helping them develop these habits, you're not just protecting their time and mental health, but you're also giving them the tools to take control of their digital lives.[119]

As we look ahead to the next three decades, the stakes will only get higher. With advancements in AI, the emotional manipulation tactics will become even more personalized and widespread. By teaching children these skills now, you're preparing them to navigate a future where the digital carnival is bigger, louder, and more tempting than ever. It's not just about surviving the digital age; it's about thriving in it while keeping their time, attention, and well-being intact.

➢ **Stay Alert to Digital Grooming and Emotional Engineering**

AI-driven grooming and manipulative conversations represent a chilling evolution in online threats. Predators no longer need to interact directly with children; they can now use AI chatbots to build false trust over time through automated conversations. These bots are designed to mimic human interaction, creating a sense of connection and familiarity that can be dangerously deceptive. For children, who may not yet have the experience to recognize these tactics, the risk is particularly high. What starts as a friendly chat with a seemingly harmless AI could slowly turn into a manipulative relationship, orchestrated by someone with malicious intent.

You must take an active role in educating your children about these risks. Teach them to recognize the signs of manipulative tactics, such as overly personal questions, requests for secrecy, or attempts to isolate them from friends and family. Emphasize that real friendships are built on trust and shared experiences over time, not through quick, intense online interactions. Encourage them to be cautious and skeptical, even when conversations feel friendly or harmless.

Reinforce the idea that no one, whether human or AI, should pressure them into sharing personal information or doing something that makes them uncomfortable. By equipping children with these tools, you're helping them build a strong defense against AI-driven grooming and other online threats. It's not just about protecting them today; it's about preparing them to navigate a future where technology continues to blur the lines between friend and foe.

As AI becomes more advanced, these risks will only grow more sophisticated. By teaching children to be vigilant and critical thinkers, you're giving them the skills they need to stay safe in an increasingly complex digital world. And while it might feel overwhelming, remember that small, consistent conversations can make a big difference.

After all, the best way to outsmart a manipulative AI is with a well-informed, confident child who knows how to spot the red flags.

Strategy #8: Reclaim Real Family Connection inside the Digital Bubble

In a world where the digital bubble often isolates family members, creating intentional moments of connection becomes a radical act of resistance. Screens may offer convenience and entertainment, but true fulfillment comes from shared human experiences.[120] Families must prioritize genuine interactions over digital distractions to maintain deep, lasting bonds. This means carving out time to be fully present with one another, whether it's through shared meals, storytelling, or simply enjoying each other's company without the interference of devices.

To foster meaningful connection, make in-person interactions non-negotiable. Schedule regular family activities that don't involve screens, like game nights, hikes, or cooking together. Encourage traditions that reinforce family identity, such as holiday rituals or weekly family meetings. These moments create a sense of belonging and remind everyone that real relationships are built on shared experiences, not digital exchanges.[121]

It's also important to limit digital replacements for real relationships. While AI companions and chatbots might seem harmless, they can never substitute the warmth and depth of human interaction.[122] Teach children that while technology can be a tool, it should never replace the irreplaceable, and that is family, friends, and the joy of being truly present with one another.

Tip #32

Raise Children to Be Digital Leaders, Not Passive Consumers

In a world dominated by algorithms and endless content, raising children to be digital leaders is more important than ever. Instead of passively consuming whatever the digital world throws their way, teach them to question and critically evaluate the content they encounter. Encourage them to think about who created it, why, and whether it aligns with their values. This critical thinking skill will help them navigate the digital landscape with confidence and discernment.[123]

Encourage creation over consumption. Whether it's writing, coding, making music, or creating digital art, these activities empower children to shape technology rather than be shaped by it. By becoming creators, they learn to see technology as a tool for expression and innovation, not just a source of entertainment.

Discuss future career opportunities in fields like AI, cybersecurity, and ethical tech development. Help them understand that the digital world isn't just something to consume, but it's something they can actively shape and improve.[124] By fostering a mindset of leadership and responsibility, you're preparing them to be pioneers in the digital age.

Tip #33

Be a Digital Guide for Your Family, Not Just a Rule Maker

Parents play a crucial role in guiding their children's relationship with technology, but this goes beyond setting rules and restrictions. Instead of simply enforcing screen time limits, strive to be a digital mentor. Model responsible technology use by demonstrating healthy habits, like putting your phone away during meals or setting aside time for screen-free activities. Children are more likely to imitate what they see than what they're told, so lead by example.

Stay updated on digital trends and the platforms your children are using. This doesn't mean you need to become an expert on every app or game, but having a basic understanding will help you guide them more effectively. Engage in open conversations about their online experiences, and create a Family Digital Agreement that establishes mutual expectations about online behavior and privacy.

By taking on the role of a mentor, you're not just protecting your children, but you're also empowering them to make smart, informed decisions about their digital lives. This approach fosters trust and collaboration, making it easier to navigate the challenges of the digital age together.

Final Reflections: Your Responsibility Toward Your Family in the Digital Bubble

In today's world, families are no longer simply using technology; they are immersed in it. By 2025, the digital bubble has become an invisible yet ever-present force shaping everyday life. From children to grandparents, every member of the household is connected. Work emails spill into dinner time. Children lose themselves in glowing screens instead of in outdoor play. AI assistants help with groceries, schedules, and even bedtime stories. While all this promises ease and efficiency, it comes at a cost that is far deeper than we often acknowledge. Beneath the surface of convenience lies a growing sense of disconnection, a weakening of the very human bonds that hold families together.

Shared laughter, mealtime conversations, and spontaneous connection once formed the heart of the home. Today, these are being displaced by isolated digital experiences. The living room, once filled with interaction, often resembles a digital waiting room where each person is immersed in their own feed or game. Parents find themselves juggling the roles of provider, protector, and tech moderator. Children are growing up in a world where identity is not rooted in experience or family values but shaped by likes, algorithms, and peer validation online. Social media magnifies insecurities, AI-generated content fills every spare moment, and real relationships begin to compete with artificially engineered ones.

Looking ahead, the influence of technology will only grow. Artificial intelligence is evolving rapidly, and it is beginning to invade some of the most sacred areas of parenting, from education and entertainment to emotional support. AI-generated friends, virtual tutors, and immersive digital environments may one day become more influential in a child's development than parents themselves. The line between human guidance and machine influence will blur. With digital profiling on the rise, the concept of privacy may be reduced to a relic of the past. If we are not vigilant, we may find our children turning to machines for companionship, validation, and advice, leaving their families as bystanders in their emotional and psychological growth.

And yet, this is not inevitable. Families still have the power to shape the role of technology in their lives. That power begins with awareness and continues with action. It means becoming intentional about how technology is used at home, setting boundaries, encouraging unplugged experiences, and resisting the passive drift into screen dependency. Establish screen-free spaces. Protect family time. Be conscious about what apps are allowed and how often devices are used. Most importantly, ensure that technology is always serving the family's well-being, not undermining it.

Digital safety must also become a shared responsibility. It is no longer enough to simply tell children not to talk to strangers online. We must equip them with the knowledge and tools to navigate a world filled with misinformation, manipulative algorithms, and AI-enhanced scams. Families must learn together how to protect their data, recognize threats, and respond to them with confidence. Teach children how to question what they see, think critically, and discern truth from manipulation. These lessons are as important today as learning how to cross the street was for earlier generations.

But perhaps the most important responsibility is to model the kind of life we hope our children will lead. When parents put down their phones, listen deeply, and spend quality time with their families, they show their children what true connection looks like. When they prioritize nature, creativity, and conversation over digital distractions, they teach their children that life outside the screen is not only real, it is richer. A family that laughs together at the dinner table, hikes in the woods, or builds something with their hands is building memories and values that no digital platform can replicate.

The future of the family in the digital bubble is not fixed. It will be defined by the choices you make now, by your willingness to reclaim attention, connection, and purpose. You do not need to reject technology. But you must refuse to be ruled by it. Families who thrive in the years to come will be those who engage with technology on their own terms thoughtfully, critically, and with human values at the center.

You are not powerless. As a parent, mentor, or guardian, your responsibility is not just to keep your family connected but to keep them grounded. The digital bubble may surround us, but it does not have to define us. Reclaim your role as the anchor in your family's digital world not by cutting cords but by strengthening the bonds that truly matter.

Chapter 8
Thriving Together — Navigating the Digital Shift as Societies and Nations

Until now we explored how individuals and families are navigating the digital era, adapting to the profound shifts brought by artificial intelligence, automation, and hyper-connectivity. While personal resilience and family structures are critical in this transformation, they exist within a larger framework shaped by society and the nation. The forces reshaping human relationships at an individual level are also redefining governance, economies, and the social fabric at scale. The digital age is not just a personal challenge; it is a collective one, where entire communities, cities, and nations must rethink their foundational structures to survive and thrive.

Societies today exist within an unprecedented digital bubble, where AI-generated realities, algorithm-driven interactions, and data-based governance are becoming the norm. The way information spreads, decisions are made, and power is exercised is shifting away from traditional human systems to AI-enhanced frameworks. While these advancements offer efficiency, innovation, and even economic growth, they also bring fragility. Trust, the foundation of any functioning society, is eroding under the weight of misinformation, deepfakes, and algorithmic manipulation. Physical communities are dissolving as digital spaces take precedence, leaving societies more connected than ever yet paradoxically more fragmented.

The challenge facing nations is no longer whether to embrace digital transformation but how to do so without sacrificing stability, fairness, and humanity. Governance powered by AI promises efficiency, but can it maintain democratic values? Economies driven by automation generate wealth, but will they also create widespread inequality? Security is shifting toward cyberwarfare and digital conflicts, raising the question of how nations will protect themselves when the next battlefield is invisible. Societies must not only contend with these immediate challenges but also prepare for the next three decades of escalating digital evolution.

Now we proceed by examining these large-scale transformations, identifying the critical challenges that societies and nations face today and the ones that will emerge in the near future. More importantly, it provides a survival guide—a roadmap for societies to ensure that they do not merely endure the digital age but actively shape it in a way that sustains human values and collective well-being. The fundamental question is whether nations will adapt and thrive or become fragmented under digital disruptions. The answer will determine the course of the century.

Part 1: The Major Challenges Societies Face in 2025

Challenge #1: The erosion of Social Trust in the Age of AI

The rapid rise of AI has created a world where telling truth from fiction is eroding. Deepfakes, AI-generated lies, and algorithm-boosted content are eroding trust in society.[125] Digital platforms, designed for engagement over truth, now spread manipulated stories and deceptive media. This lets bad actors distort facts, sway public opinion, and erode confidence in institutions. Despite being more connected digitally, people now distrust the information they see and the systems that guide them.

Beyond digital deception, declining in-person interactions are eroding community ties. As life moves online, local events, civic groups, and casual social moments are vanishing. The consequences are clear: less empathy, weaker relationships, and growing isolation. Some now prefer AI companions over real conversations, eroding their connection to the physical world.[126] Without intervention, we may lose the shared experiences that once built trust and cooperation.

The critical question remains: Can we rebuild trust when reality itself is eroding? If AI advances unchecked, the line between truth and lies may disappear entirely. Without action, collapsing trust could destabilize society, making democracy, governance, and daily life increasingly fragile. The challenge is enormous, but the cost of inaction is higher.

Challenge #2: Polarization & algorithmic bias—Digital bubble reinforces societal divides

Digital algorithms trap users in echo chambers by feeding them content that matches their existing views.[127] This reinforcement cycle widens societal divides, as people only see information that confirms their biases while alternative perspectives get silenced. The result? Rising polarization that fractures communities.

The damage runs deep. Instead of fostering healthy debate, these systems magnify division, giving extreme ideas fertile ground to grow. Someone with mild political leanings might be fed increasingly radical content, pushing them to ideological extremes. This destroys the shared understanding needed for cooperation and turns disagreements into unbridgeable chasms.

Political operatives exploit these systems as weapons. They craft tailored messages to motivate groups against each other, plant misinformation, and dismantle trust in institutions.[128] The outcome is a broken society where people see opponents as threats rather than neighbors with differing views.

The crisis accelerates because lies spread unchecked online. One viral post can poison millions of minds before fact-checkers respond. This instant contamination breeds societies fractured by suspicion and hostility.

Without intervention, the consequences will be permanent. Imagine communities retreating into ideological bunkers, colleagues refusing to collaborate across party lines, and the social contract crumbling. Fixing this requires rebuilding social media from the ground up with truth and unity as core priorities.

Challenge #3 Decline of physical communities

As digital technologies continue to dominate our lives, traditional forms of social interaction are increasingly being replaced by virtual engagement. The shift from face-to-face communication to online interactions has profound implications for the fabric of society. Physical communities, once the cornerstone of social cohesion, are weakening as people spend more time in digital spaces. This decline in physical community engagement has far-reaching consequences, including reduced empathy, weaker social bonds, and a growing sense of isolation.

One of the most significant impacts of this shift is the erosion of empathy. In-person interactions allow for nuanced communication, where body language, tone, and facial expressions play a crucial role in understanding others. Virtual interactions, on the other hand, often lack these nuances, making it easier for misunderstandings to arise and for empathy to diminish. Over time, this can lead to a more fragmented society where individuals feel less connected to one another.

The decline of physical communities also weakens social bonds. Traditional communal structures, such as local gatherings, civic organizations, and even casual social encounters, are diminishing as people turn to digital platforms for connection. While online communities can provide a sense of belonging, they often lack the depth and resilience of real-world relationships. This weakening of social bonds can lead to increased feelings of loneliness and isolation, which have been linked to a range of mental health issues.

Moreover, the rise of AI-generated influencers and digital avatars is further detaching individuals from the physical world. These virtual entities, designed to mimic human interaction, can create the illusion of connection without the substance of real relationships. As people spend more time engaging with these digital counterparts, the risk of losing touch with genuine human connections grows.

The consequences of this decline in physical communities are profound. Reduced empathy, weaker social bonds, and increased isolation can lead to a more fragmented and less cooperative society. If left unchecked, the shift toward virtual engagement could undermine the very essence of human cooperation that has historically been the foundation of progress. Addressing this challenge requires an effort to rebuild and strengthen physical communities, ensuring that they remain a vital part of our social fabric in the digital age.

Challenge #4: AI-Led Governance Balancing Data-Driven Decisions with Democratic Values

The growing use of AI in government creates a tricky situation. While AI promises to make policies more efficient and data-driven, it also threatens democratic oversight. Governments now use AI for everything from city planning to criminal justice, relying on its ability to process huge amounts of information and spot patterns.[129] But these systems have serious flaws that could damage public trust.

A major problem is that AI decisions today are often impossible to explain. Many systems work like secret machines where even experts can't fully understand how they reach conclusions. When AI determines who gets social benefits or who is denied a bank loan, people can't question the results because no one can properly explain them. This makes government less answerable to citizens.

AI also carries hidden biases. Since these systems learn from past data, they often repeat old mistakes and discrimination. A policing AI trained on unfair arrest records will keep targeting the same neighborhoods. A benefits system trained on biased data might wrongfully deny help. Instead of fixing inequalities, AI can make them worse while giving them a false appearance of scientific fairness.

Democracy suffers when algorithms make important choices. Normally, elected leaders must explain their decisions and face public criticism. But when AI calls the shots, no one takes responsibility. Voters can't challenge software the way they can challenge politicians. This makes people feel powerless and disconnected from their own government.

The situation also gives too much power to tech companies. Private firms design these AI systems but face little supervision. Their hidden algorithms shape lives while avoiding democratic checks. Ordinary citizens lose say in how they're governed as unelected corporations gain influence over policy decisions.

Left unchecked, AI governance could create efficient but heartless systems that value data over people. Algorithms might optimize cities for statistics rather than livability or distribute resources based on calculations rather than justice. The convenience of AI comes with the risk of weakening the public's control over the decisions that affect them most.

Challenge #5: Surveillance states

The rapid expansion of AI-powered surveillance technologies has created a delicate and often shaky balance between enhancing security and violating individual freedoms. Governments across the globe are increasingly implementing AI-driven monitoring systems, ranging from facial recognition networks to predictive policing algorithms, under the pretext of public safety and national security. While these innovations offer law enforcement new tools to prevent crime and maintain order, they also introduce profound risks to privacy, civil liberties, and democratic principles. The distinction between legitimate security measures and an overreaching surveillance state is becoming increasingly blurred, raising critical ethical and societal concerns.

One of the most pressing threats is the erosion of privacy. AI-driven surveillance systems continuously collect, process, and store vast amounts of personal data, often without individuals' knowledge or consent. Facial recognition technology can track people's movements in real time,[130] while sophisticated AI algorithms analyze online activity to predict behavior. Such monitoring creates an environment where citizens are constantly watched, creating an atmosphere of fear that suppresses free expression and discourages opposition. Fear of surveillance may push individuals to self-censor or avoid certain activities, ultimately weakening the core freedoms upon which democratic societies are built.

The potential for misuse of these surveillance tools is another major concern. In authoritarian regimes, AI-powered monitoring can become a means of oppression, allowing governments to silence political opposition and exert control over populations. Even in democratic nations, the improper use of surveillance data by authorities can result in discrimination and human rights violations. Predictive policing algorithms, for instance, have been widely criticized for disproportionately targeting marginalized communities, reinforcing existing biases, and deepening social inequalities.[131]

Beyond these immediate concerns, the normalization of mass surveillance threatens to reshape societal norms, leading to gradual acceptance of constant monitoring as an unavoidable reality. As citizens grow indifferent to being watched, governments may gain unprecedented control over their populations, getting closer to digital authoritarianism. The once-dystopian concept of an all-seeing "Big Brother" is no longer confined to fiction; in many parts of the world, it is already becoming a tangible reality.

The rise of surveillance states could lead to a future where personal freedoms are systematically sacrificed in the name of security, granting governments excessive control over their citizens. To address this growing challenge, societies must strike a careful balance between utilizing surveillance technologies for public safety and safeguarding fundamental rights.

Part 2: The Challenges That Will Arise Over the Next 3 Decades

Challenge #1: Mass job displacement

The next three decades will witness an unprecedented shift in the global workforce as artificial intelligence and automation redefine entire industries.[132] While technological advancements have historically created new opportunities alongside disruptions, the scale and speed of AI-driven transformation pose a unique challenge. Unlike past industrial revolutions, where machines primarily replaced physical labor, AI and robotics are now capable of performing cognitive tasks, decision-making processes, and even creative functions. As automation becomes more sophisticated, millions of jobs across various sectors are at risk of being eliminated or radically transformed.

Industries reliant on repetitive, rule-based tasks will be the first to experience large-scale job displacement. Manufacturing, already heavily automated, will continue to see a decline in human labor as intelligent robotics take over assembly lines, quality control, and logistics. Similarly, transportation and delivery services will be reshaped by autonomous vehicles, reducing the demand for drivers in trucking, ride-hailing, and public transport. Retail and customer service roles are also vulnerable, as AI-powered chatbots, self-checkout systems, and robotic assistants gradually replace human workers.

Beyond blue-collar industries, AI is poised to disrupt white-collar professions once considered immune to automation.[133] The rise of generative AI and advanced algorithms threatens jobs in finance, law, healthcare, and even creative fields. For instance, AI-powered financial analysis tools can process vast amounts of data faster than human analysts, while legal AI systems can draft contracts, conduct research, and predict case outcomes with remarkable accuracy. In healthcare, AI-driven diagnostics and robotic surgeries are reducing the need for certain medical professionals. Even artistic and content-driven industries are not exempt, as AI-generated music, writing, and visual art become increasingly sophisticated.

The scale of job displacement raises fundamental questions about economic stability, social inequality, and the future role of human labor. As automation eliminates jobs at a faster rate than new industries can emerge, governments and businesses must proactively address workforce transitions. Without strategic policies and investment in reskilling programs, millions of workers could be left unemployed, widening the economic divide and exacerbating social unrest.

The challenge is not just about job loss but about redefining work itself. Will societies successfully adapt by creating new employment opportunities, or will AI-driven economies leave large portions of the population struggling to find meaningful work? Navigating this transition will require a collaborative effort between governments, businesses, and educational institutions to prepare workers for an era in which adaptability and lifelong learning become essential survival skills. The choices made in the coming decades will determine whether AI leads to a more prosperous, inclusive society or one defined by widespread economic displacement and deepening inequality.

Challenge #2: AI-driven economic inequality

As artificial intelligence becomes a key driver of economic growth, a new era of wealth concentration is taking shape. Prosperity is increasingly controlled by a small group of powerful tech companies and individuals. Over the next 30 years, automation, data control, and algorithm-based decision-making will reshape how wealth is distributed. Those who create and own AI systems will likely gain more power and income, while those working in traditional jobs may face greater insecurity. This growing divide could lead to economic instability, social tension, and major shifts in how power is shared between countries, corporations, and everyday people.

One of the main reasons for this inequality is that economic power is becoming concentrated in tech-focused industries. The companies leading the AI wave—from cloud services and automation to biotech and finance—are gaining more wealth and control than ever before. They have access to massive amounts of data, cutting-edge computing tools, and exclusive rights to key technologies. This gives them a huge advantage and makes it hard for smaller competitors to keep up. As AI makes businesses more efficient, traditional industries that do not adapt may fall behind, leading to job losses and economic disruption.

At the same time, AI-driven automation disproportionately benefits capital over labor. Unlike previous industrial revolutions, where technological advancements increased the demand for workers in new fields, AI is replacing human labor in both manual and cognitive jobs at an accelerating pace.[134] This shift means that wealth generated through automation is flowing primarily to business owners, investors, and AI developers, rather than being distributed across the workforce. The result is an economy where a shrinking elite collects the financial rewards while the majority of workers face declining wages, underemployment, or even job displacement.

Beyond corporate power, AI-driven inequality also extends to national economies, increasing the divide between technologically advanced nations and those struggling to compete. Countries with strong AI research, access to big data, and significant investment in automation are poised to dominate global markets, leaving developing nations at a disadvantage.[135] This digital divide threatens to create a two-tiered world economy, where nations that fail to integrate AI into their economic models experience decline and dependency on AI-rich countries for innovation and economic growth.

The central challenge for policymakers and global institutions is finding ways to ensure that AI-driven prosperity benefits society as a whole, rather than deepening economic divisions. Potential solutions include implementing progressive taxation on

AI-driven industries, redistributing wealth through universal basic income or alternative social safety nets, and investing heavily in reskilling programs to help workers transition into AI-enhanced economies. Without strategic intervention, the unchecked concentration of AI-generated wealth could lead to extreme economic stratification, eroding social cohesion and fueling political instability.

In the coming decades, societies will either find ways to use AI to create fair economic opportunities for everyone or risk a future where only a small group of people who own and control AI technology hold most of the wealth and power.

Challenge #3: The Race for Digital Dominance

If the Cold War was a battle over nuclear power, the next great global rivalry is all about artificial intelligence and quantum computing. Instead of missiles, nations are stockpiling algorithms. Instead of spies sneaking into enemy territory, governments are deploying AI-driven cyberattacks and quantum encryption breakers. And instead of an arms race, we now have an "AI-Quantum race" where the winner doesn't just gain technological prestige but also economic power, military superiority, and the ability to rewrite the rules of the digital world.

The competition is intense. Superpowers like the U.S. and China are leading the charge, pouring billions into AI research and racing to achieve quantum supremacy, which is the point where quantum computers can outclass even the most advanced classical supercomputers.[136] Meanwhile, smaller nations are scrambling to keep up, hoping they don't end up as "digitally dependent states" under AI-quantum superpowers. Every country wants to dominate these fields because whoever controls the best AI and quantum technology will dictate global markets, cybersecurity, and even real-time manipulation of financial systems and intelligence networks.

But this isn't just about raw computing power; it's about strategic maneuvering. Nations are engaging in high-stakes digital espionage, hacking into rivals' AI and quantum labs, and hoarding vast amounts of data like it's the new gold. Some are forging AI alliances, while others are setting up strict regulations to prevent foreign tech giants from gaining too much control. Meanwhile, corporations like *Google*, *IBM*, *OpenAI*, and *Huawei* are acting like digital superpowers of their own, wielding more influence than some governments.

The real question is whether this AI-Quantum competition will lead to breakthroughs that uplift humanity or a chaotic struggle for control, where ethical considerations are tossed aside in favor of digital dominance. Will nations collaborate to ensure AI and quantum computing are developed responsibly, or will we see the rise of quantum-powered cyberwars, AI-controlled propaganda machines, and a new era of algorithmic imperialism?

One thing is certain: this race is accelerating, and whoever wins won't just shape the future of AI and quantum computing; they'll shape the future of civilization itself.

Challenge #4: Cyberattacks on critical infrastructure

As societies become increasingly digital, their most essential systems, such as energy grids, financial networks, and national defense infrastructures, are more vulnerable than ever to cyberattacks. What once required armies and physical invasions can now be achieved with a few keystrokes and well-designed malware. In the coming decades, cyber warfare will not just be a side concern; it will be a primary battleground where nations, criminal organizations, and rogue actors compete for power, money, and control.

Energy infrastructure is one of the most high-risk targets.[137] A successful cyberattack on an electrical grid could plunge entire cities into darkness, disrupt industries, and bring daily life to a standstill. Hackers with access to AI-powered malware could sabotage power plants, manipulate energy markets, or even hold entire nations hostage by demanding ransom in exchange for restoring operations. With the rise of smart grids and interconnected energy systems, the attack surface is only expanding, making it increasingly difficult to secure every access point.

The financial sector is another prime target. AI-enhanced cyberattacks could manipulate stock markets, cripple banking networks, or drain billions from accounts before security systems even detect the breach. As digital currencies and blockchain technologies become more integrated into financial systems, new vulnerabilities will emerge, offering hackers sophisticated ways to exploit the economy. A well-coordinated cyber assault on major financial institutions could trigger market crashes, disrupt trade, and erode trust in global banking systems.

National defense is no longer just about missiles and military forces; it is about protecting digital command centers, satellite networks, and AI-driven warfare systems. Cyberattacks on defense infrastructure could disable critical military communications, compromise classified intelligence, or even hijack autonomous weapons. Nations that fail to secure their defense networks risk having their own systems turned against them. With AI-driven cyber warfare becoming more advanced, it is no longer just about protecting against human hackers but also about countering AI algorithms that can adapt, learn, and exploit vulnerabilities faster than any human can respond.

As cyberattacks grow more sophisticated, the biggest challenge will be staying one step ahead. Governments, corporations, and institutions must invest heavily in cybersecurity, build resilient infrastructures, and develop AI-driven defense mechanisms capable of countering AI-driven threats. The coming decades will test whether societies can safeguard their digital foundations or if they will fall victim to the next generation of cyber warfare. The choice is clear: either strengthen digital defenses or risk watching critical systems crumble at the hands of unseen enemies.

Challenge #5: Autonomous weapons & AI-driven conflicts

Warfare is undergoing a seismic transformation, moving far beyond the traditional domains of rockets, tanks, and battlefields. The next generation of military power will be defined by artificial intelligence, autonomous weapons, and algorithm-driven strategies. Nations are already pouring vast resources into AI-enhanced warfare, developing technologies such as drones capable of striking without human intervention, robotic soldiers that operate tirelessly, and cyber systems designed to cripple an enemy's infrastructure before a single bullet is fired. As AI continues to evolve, the question is no longer whether it will change warfare. It already has. The real question is whether humans will retain any meaningful control over the battlefield.

One of the most alarming developments is the rise of fully autonomous weapons, often referred to as killer robots.[138] These AI-powered systems can identify, target, and eliminate threats without human oversight, promising unprecedented speed, precision, and reduced human casualties for the side deploying them. However, they also introduce profound ethical and strategic dilemmas. Unlike human soldiers, an AI drone does not experience fear, hesitation, or moral judgment. If programmed with flawed logic or biased algorithms, it could misidentify targets, escalate conflicts unintentionally, or even be hacked and turned against its own operators. The prospect of machines making life-and-death decisions raises fundamental questions about accountability, morality, and the very nature of warfare.

Adding to these challenges is the global AI arms race. As one nation develops advanced autonomous weapons, its rivals feel compelled to do the same, creating a dangerous cycle of escalation. Without international agreements or ethical guidelines, the risk of unintended wars increases dramatically. History has shown that arms races lead to instability, and when AI is involved, the speed at which decisions are made could outpace human ability to intervene in time to prevent catastrophe. The stakes are even higher when machines are entrusted with decisions that could trigger global conflicts.

The future of military power hinges on whether nations can establish clear rules and boundaries for the use of AI in warfare. Will world leaders agree to limit the deployment of autonomous weapons, or will they embrace an era where wars are fought by machines, with humans relegated to the sidelines? The coming decades will determine whether AI is harnessed to enhance security and defense or if it will lead to a world where battles are decided by cold, calculating algorithms. The prospect of machines making irreversible, life-and-death decisions should give humanity pause. But whether that pause translates into meaningful action remains to be seen. The choices made today will shape not only the future of warfare but also the ethical and moral foundations of global security.

Part 3: How Societies Can Not Just Survive, But Thrive in the digital bubble

After navigating the minefield of deepfakes, job automation, cyber warfare, AI-driven inequality, and digital surveillance, one thing is clear: societies cannot afford to sit back and hope for the best. The digital revolution is not slowing down, and it is not just about artificial intelligence. From blockchain to quantum computing, from social media to biotech, digital technologies are rewriting the rules of how societies function. The real question is not whether these technologies will shape the future, but how societies can shape them in a way that preserves human values, economic stability, and democratic principles.

Every major technological revolution has brought both chaos and opportunity. The Industrial Revolution disrupted traditional economies but also lifted millions out of poverty.[139] The digital era will be no different. Societies that learn to harness technology while maintaining human agency will lead the future. Those that fail to do so will find themselves at the mercy of a world run by algorithms, monopolies, and cyber vulnerabilities.

The challenge ahead is not about resisting digital transformation, nor is it about blindly embracing every new technology. It is about finding balance. How can nations protect jobs while embracing automation? Can governments ensure security without turning into surveillance states? Will societies create inclusive digital economies, or will wealth and power be concentrated in the hands of a tech elite? These are not futuristic hypotheticals; they are urgent questions that demand action now.

This section is about solutions. It is about taking control of the digital era instead of letting it control us. It is about policies that foster innovation without sacrificing privacy, education systems that prepare citizens for a digital-first world, and economic models that ensure prosperity is shared, not hoarded. The goal is not just to survive the digital bubble but to thrive in it by leveraging technology to create smarter, fairer, and more resilient societies. The future is unwritten, and the next few decades will determine whether digital technologies become tools for collective progress or forces of division and control. The choice is ours to make, and I hope that we choose wisely.

Strategy #9: Set Digital literacy as a national priority

As digital technologies continue to evolve at breakneck speed, societies that don't keep up risk becoming the digital equivalent of a pager in a smartphone world. Economically and politically, the stakes are high. Digital literacy isn't just about knowing how to turn on a computer or scroll through social media anymore. It's about understanding how artificial intelligence, big data, blockchain, and cybersecurity shape everything from your morning coffee order to global governance. Societies that prioritize digital literacy won't just protect themselves from manipulation and misinformation—they'll also be the ones calling the shots in the global digital economy.

One of the biggest threats to modern societies is the rapid spread of misinformation and AI-generated trickery. Deepfakes, algorithmic biases, and data-driven manipulation can destabilize democracies, sway elections, and turn social media into a battleground. Without a digitally literate population, societies risk becoming sitting ducks for bad actors, both foreign and domestic, who can control narratives and erode trust in institutions. Teaching people how to critically assess online content, spot AI-generated media, and verify sources is as essential today as teaching children to read and write was a century ago. After all, if you can't tell the difference between a real news story and a deepfake of a politician singing karaoke, you're in trouble.

But digital literacy isn't just about defense. It's also the golden ticket to economic resilience. The job market of the future will be dominated by automation, AI, and data-driven industries. Countries that prepare their workforce with AI education, coding skills, and data competency will thrive, while those that ignore these shifts will be left wondering why their economy is stuck in the dial-up era.[140] The demand for AI-savvy professionals, cybersecurity experts, and digital entrepreneurs is skyrocketing. Without strategic investment in digital education, societies risk creating a massive skills gap, where a small tech-savvy elite gains the benefits of new technologies while everyone else is left Googling "what is blockchain?"

Take Estonia, for example. This forward-thinking Baltic nation has always been a step ahead when it comes to embracing the digital future. In 2025, Estonia launched an ambitious nationwide artificial intelligence education program called *AI Leap 2025* (or *TI-Hüpe 2025* in Estonian).[141] This groundbreaking initiative, set to roll out on September 1, 2025, provides students and teachers with free access to the world's leading AI applications and equips them with the skills to use these tools effectively in learning. But Estonia didn't stop at schools—adults were also given access to AI-focused upskilling programs, ensuring the entire workforce remained competitive in a rapidly evolving job market.

By making digital competency a national priority, Estonia has transformed itself into one of the most technologically advanced and resilient societies in the world. It's as if they saw the digital revolution on the horizon and said, "We'll take the front row, thanks." This proactive approach not only prepares the next generation for an AI-driven future but also ensures that the current workforce isn't left behind. Estonia's commitment to digital literacy and AI education is an example of how societies can thrive in the digital age by embracing innovation and equipping their citizens with the tools to succeed.

Governments need to integrate AI and digital literacy into their education systems, ensuring that young people are prepared for a digital-first world. Adult learning programs should be expanded to help workers transition into AI-driven industries, and national campaigns should promote cyber awareness, fact-checking, and responsible online behavior.[142] Think of it as a digital survival kit for the 21st century.

The future will belong to societies that don't just passively consume technology but actively shape it. The nations that treat digital literacy as a core pillar of their development strategy won't just defend themselves against digital threats—they'll also build stronger economies, more informed citizens, and a workforce ready for the AI age. The challenge is clear. Will societies embrace digital literacy as a priority, or will they let technological progress leave them in the dust? The next three decades will reveal the answer, and the clock is ticking. So, grab your digital toolkit—it's time to get literate.

The future will belong to societies that do more than passively consume technology. They will actively shape it. Nations that treat digital literacy as a core pillar of their development strategy will not only defend themselves against digital threats but also build stronger economies, more informed citizens, and a workforce ready for the AI age. The challenge is clear. Societies must embrace digital literacy as a priority or risk being left behind by the relentless pace of technological progress. The next three decades will reveal which nations rise to the occasion and which fall by the wayside. The clock is ticking, and the time to act is now. Grab your digital toolkit. It's time to get literate.

Strategy #10: Rebuild human-centric policies

For societies to thrive in the digital age, artificial intelligence must be developed and implemented with a clear focus on enhancing human potential rather than replacing it. The future cannot be dictated by algorithms alone. Governments and industries must adopt policies that ensure AI serves as a collaborative tool, strengthening human capabilities rather than diminishing them. A human-centric approach to AI is essential to creating a future where technology supports well-being, economic stability, and social cohesion.

A key aspect of this strategy is redesigning workplace automation to complement human skills instead of making them obsolete. AI-driven systems should be integrated in ways that free workers from repetitive, routine tasks while allowing them to focus on creativity, critical thinking, and interpersonal collaboration. Governments must establish regulations that encourage businesses to implement AI augmentation models, where human expertise remains central and technology acts as a support mechanism rather than a replacement. Reskilling and upskilling programs should be a national priority, ensuring that workers can transition smoothly into roles where AI enhances productivity rather than eliminates jobs.

Beyond the workplace, AI-driven decision-making must include built-in human oversight to prevent biases, errors, and ethical blind spots. Governments should enforce transparency in AI policies, requiring clear accountability mechanisms for automated systems used in sectors like healthcare, law enforcement, and governance. AI tools should amplify fairness and accessibility, ensuring that digital transformation does not deepen social and economic divides. Regulatory frameworks should mandate explainable AI, where individuals have the right to understand and challenge algorithmic decisions that impact their lives.

Human-centric AI policies also need to focus on public well-being. AI should improve healthcare accessibility, education quality, and government services, making daily life more efficient while ensuring that individuals still have agency over personal choices. Governments must push for AI ethics guidelines that prioritize human rights, prevent mass surveillance abuse, and ensure AI is developed with inclusivity in mind.

The societies that embrace AI as a partner rather than a replacement will be the ones that flourish in the next three decades. By placing human needs, creativity, and ethical responsibility at the core of AI policies, nations can harness the benefits of digital transformation while preserving economic security, social trust, and personal autonomy. A future where AI enhances life rather than dictates it is possible, but only if societies make deliberate choices today to build a human-first digital era.

Strategy #11: Cooperate on a Global AI Governance

The idea of nations and corporations coming together to establish a fair, transparent, and ethical framework for AI governance sounds like the plot of a utopian sci-fi movie. It's the kind of thing you'd expect to see in a world where everyone holds hands and sings kumbaya while algorithms hum in the background. Yet, history has shown that when the stakes are high enough, global cooperation isn't just possible, it's necessary. The world has managed to agree on nuclear weapons, climate change, and international trade, despite everyone's love for arguing over who gets the biggest slice of the pie. The challenge with AI governance, though, is that the forces driving its development are more interested in dominance, profits, and strategic advantage than in playing nice.

AI isn't just another shiny new gadget. It's a power multiplier on steroids. The nations that lead in AI development will control economic systems, cybersecurity, military strategy, and even public perception through AI-driven media manipulation. This creates the risk of digital imperialism, where a few AI superpowers call the shots while everyone else scrambles to catch up. Without a global AI framework, the world could end up with a digital divide so deep that the powerful control not just AI but also access to information, economic opportunities, and security systems. It's like a high-stakes game of Monopoly, except the board is the planet, and the rules are written by whoever owns the most data.

But global AI governance isn't entirely a pipe dream. The world has already pulled off some pretty impressive cooperations, like the Nuclear Non-Proliferation Treaty (NTP), which 191 countries signed because, well, nobody wants to wake up to a mushroom cloud. AI, while not as explosive, poses similarly unpredictable and far-reaching risks. An AI arms race could lead to autonomous weapons that make decisions faster than humans can blink, economic warfare through algorithmic manipulation, or a surveillance state that makes Big Brother look like a friendly neighbor. If nuclear weapons needed containment, AI definitely needs some ground rules before it spirals out of control.

The real kicker, though, is corporate power. Unlike nuclear weapons, AI isn't just in the hands of governments. It's largely developed by private companies whose primary goal is profit, not global stability. Tech giants operate across borders, hoarding data, refining algorithms, and consolidating power with all the oversight of a toddler in a candy store. Expecting corporations to self-regulate is like expecting a cat to guard a fishbowl. Without strong international policies, AI-driven monopolies will shape the world based on market interests, not ethical considerations.

To avoid a future where AI is controlled by a handful of tech overlords, global cooperation needs to happen, in my opinion, at two levels. First, nations need to create a binding international agreement on AI governance, ensuring transparency, ethical use, and limits on AI warfare. Second, governments must enforce corporate accountability, preventing AI from being hijacked by private entities that care more about profits than people. Without these safeguards, the world risks a future where AI serves the interests of the few while the rest of us deal with economic exclusion, digital surveillance, and algorithmic oppression.

To manage the risks of AI, we need a strong international agreement—a Development and Control Treaty—to set clear rules.[143] This treaty would stop the unchecked spread of powerful AI systems like artificial general intelligence.[144] It would ensure that AI is used for good, such as helping with medical research, and not for harmful uses like autonomous weapons. The treaty would include strict checks to track how AI is developed, including where the data and computing power come from. Like the Nuclear Non-Proliferation Treaty, only trusted government and scientific groups would be allowed to work with the most advanced AI systems, and their work would be regularly reviewed.[145] A separate Test Ban Treaty would stop any AI system from being used in the real world until it has passed strong safety checks, similar to how nuclear tests are limited to controlled settings.[146]

We also need similar protections for quantum technology. A Quantum Safeguard Pact would ban using quantum computers for dangerous purposes, like breaking encryption or targeting weapons. It would limit who has access to sensitive parts like qubit processors while still allowing the use of quantum tools in helpful areas like materials science. This kind of agreement would work much like the Missile Technology Control Regime, which helped stop the spread of dangerous missile systems.

To go further, an Espionage Agreement would ban countries from stealing cutting-edge research. In the quantum age, stealing a breakthrough could be just as harmful as stealing nuclear secrets once was.

Together, these proposals apply lessons from arms control to the modern era. Both AI and quantum tech have the potential to either greatly help humanity or cause serious harm. Like past treaties that reduced nuclear threats through oversight and cooperation, these new agreements would create global rules for transparency, safety, and ethical use, protecting the future while still supporting positive scientific progress.

The next three decades will determine whether technology governance becomes a collaborative effort for the benefit of humanity or a fragmented struggle for control. While artificial intelligence is at the forefront of this challenge, other emerging technologies such as quantum computing and advanced automation will also require global oversight. If nations have been able to negotiate treaties on weapons capable of mass destruction, there is no reason they should not be able to establish agreements on technologies that have the potential to reshape the world. The real issue is whether governments and corporations will take action before the risks become irreversible or if they will delay until these technologies have already altered society beyond repair. Time is running out, and the consequences could not be more serious. With so much at stake, it is crucial that leaders act before reality starts to resemble the plot of a dystopian blockbuster.

Strategy #12: Balance Technological Progress with Social Well-Being

The key to maintaining balance is to create robust regulatory frameworks that encourage ethical and responsible technology development. Governments must develop policies to ensure that AI and other digital breakthroughs benefit the public rather than just business or political interests. This entails enacting explicit legislative safeguards to prevent misuse, promoting transparency in decision-making, and enforcing accountability from both governments and technology corporations. Without these safeguards, technological advancement risks becoming an instrument for control and exploitation rather than a force for societal benefit.

The public's awareness of AI and digital technology is as vital as regulation. Societies cannot rely exclusively on governments to navigate this transformation. Promoting digital literacy allows people to make educated decisions about how technology affects their lives. Governments, educational institutions, and media organizations must collaborate to guarantee that talks about AI and digital governance are open to everyone, not just specialists and industry leaders.

Another essential component of this strategy is to promote inclusive decision-making processes that prioritize the common good. The development and implementation of new technologies should not be influenced only by corporate interests or a few powerful countries. Public discourse, civic involvement, and international cooperation must set the ethical boundaries of AI and digital tools. Without participation from various populations, technology risks perpetuating biases, increasing economic inequality, and eroding democratic norms.

A sustainable future necessitates that security does not come at the expense of liberty. Surveillance technology, prediction algorithms, and automated governance systems must be monitored and regulated to avoid overreach. Strong supervision systems are required to ensure that AI is not used for mass surveillance or societal control. Active civic involvement is critical for holding governments and corporations accountable and ensuring that privacy, human rights, and social well-being remain at the heart of technological innovation.[147]

The next three decades will determine whether society can incorporate AI and other digital breakthroughs in ways that benefit humanity rather than damage it. The task is to guide change in a way that values fairness, decency, and shared wealth. With responsible governance, informed public discourse, and ethical leadership, societies can ensure that technological advancement supports, not diminishes, the fabric of human civilization.

Final Reflections: Our Shared Responsibility in the Digital Society

The digital revolution has come, not only changing our means of communication and employment but also our society's most basic functioning. It is no longer approaching. Our ability to face difficult issues, question accepted wisdom, and act boldly will determine how digital-era civilization turns out. Nations have to be stewards of human well-being in an era defined by algorithms and automation in addition to being controllers of technology.

Right now, we live at a pivotal junction in history. Decisions taken over the next three decades will decide whether technology becomes an instrument for common progress or a factor worsening inequality, endangering democracy, and tearing the social fabric. If not directed by strong, ethical governance, artificial intelligence, automation, quantum computing, and surveillance capabilities run the danger of worsening already existing cracks both inside and across civilizations. The stakes cannot be higher. No one organization bears sole responsibility for guaranteeing a good future. Governments have to create forward-looking rules that guarantee technology helps people instead of substitutes or controls them. Companies have to understand that profit cannot come before trust from the public, equity, or security. Moreover, as citizens, we have to actively participate in forming the digital rules and policies controlling our lives. Unless we view civic participation, openness, and responsibility as non-negotiable, democracy itself will not last the digital age.

This is a moral and civic as well as a technical struggle. Making sure the digital era enhances rather than compromises society calls for us to give empathy first priority over efficiency, openness over convenience, and cooperation over competitiveness. We have to rediscover what it means to be human as societies adjust to fast technology development and make sure our systems mirror those values.

The good news is that progress's tools are within grasp. Investing in digital literacy, supporting ethical AI development, safeguarding civil liberties, and promoting worldwide collaboration can help nations orient toward a more fair, strong, and inclusive digital future. Societies may not only survive but thrive inside the digital bubble by giving human well-being top priority, building protections, and encouraging innovation with intent.

The future is not something to run from or mindlessly welcome. We can shape something together. The argument is no longer whether we live in a digital bubble. What type of civilization we wish to create inside this new reality is the fundamental concern. The time to answer that question is now.

Stay Connected

As you step back from these pages and return to your own digital world, I invite you to carry this question with you: What kind of human do I want to be in a technology-shaped future? If this book has sparked reflection, challenged your thinking, or offered clarity, I would truly love to hear from you. Visit TamerBadawy.com to share your thoughts, ask a question, or explore how we can shape a more human future together. You will also find practical tools, fresh updates, and meaningful ways to stay connected and engaged beyond the book. Let this be the beginning of an ongoing conversation.

Warm regards,

Tamer Badawy

Villars-sur-Glâne, 2025

Endnotes

[1] David McNeill, *Hand and Mind: What Gestures Reveal about Thought* (Chicago, IL: University of Chicago Press, 1996).

[2] Andrew Robinson, *The Story of Writing* (London: Thames & Hudson, 2007).

[3] Elizabeth L. Eisenstein, *The Printing Revolution in Early Modern Europe* (Cambridge University Press, 2005).

[4] Tom Standage, *The Victorian Internet: The Remarkable Story of the Telegraph and the Nineteenth Century's On-Line Pioneers* (Walker and Company, 1998).

[5] Virginia Danielson, *"The Voice of Egypt": Umm Kulthum, Arabic Song, and Egyptian Society in the Twentieth Century*, Chicago Studies in Ethnomusicology (Chicago, IL: University of Chicago Press, 1998).

[6] Janet Abbate, *Inventing the Internet* (Cambridge, Mass.: MIT Press, 2000).

[7] Averroës, *Decisive Treatise and Epistle Dedicatory*, trans. Charles E. Butterworth (Provo, Utah: Brigham Young University, 2001).

[8] Abbate, *Inventing the Internet*.

[9] Walter Isaacson, *The Innovators: How a Group of Hackers, Geniuses, and Geeks Created the Digital Revolution* (New York: Simon & Schuster, 2015).

[10] Brad Stone, *The Everything Store: Jeff Bezos and the Age of Amazon* (New York: Little, Brown and Company, 2013).

[11] John Cassidy, *Dot.Con: The Greatest Story Ever Sold* (New York: Harper, 2002).

[12] Fred Vogelstein, *Dogfight: How Apple and Google Went to War and Started a Revolution* (New York: Sarah Crichton Books, 2013).

[13] Clay Shirky, "The Political Power of Social Media: Technology, the Public Sphere, and Political Change," *Foreign Affairs* 90, no. 1 (2011): 28–41.

[14] Safiya Umoja Noble, *Algorithms of Oppression: How Search Engines Reinforce Racism* (NYU Press, 2018).

[15] Samuel Greengard, *The Internet of Things* (Cambridge, Massachusetts: The MIT Press, 2015).

[16] Kevin Ashton, *How to Fly a Horse: The Secret History of Creation, Invention, and Discovery* (New York: Anchor, 2015).

[17] Shoshana Zuboff, *The Age of Surveillance Capitalism: The Fight for a Human Future at the New Frontier of Power* (New York: PublicAffairs, 2019).

[18] Kai-Fu Lee, *AI Superpowers: China, Silicon Valley, and the New World Order* (Boston, MA; New York, NY: Harper Business, 2021).

[19] Max Tegmark, *Life 3.0: Being Human in the Age of Artificial Intelligence* (New York: Knopf, 2017).

[20] Ray Kurzweil, *The Singularity Is Near: When Humans Transcend Biology* (East Rutherford: The Viking Press, 2005).

[21] Adam Greenfield, *Radical Technologies: The Design of Everyday Life* (London: Verso, 2017).

[22] Zuboff, *The Age of Surveillance Capitalism*.

[23] Cathy O'Neil, *Weapons of Math Destruction: How Big Data Increases Inequality and Threatens Democracy* (New York: Crown, 2017).

[24] Eli Pariser, *The Filter Bubble: How the New Personalized Web Is Changing What We Read and How We Think* (London: Penguin Books, 2012).

[25] Erik Brynjolfsson and Andrew McAfee, *The Second Machine Age: Work, Progress, and Prosperity in a Time of Brilliant Technologies* (New York; London: W. W. Norton & Company, 2016).

[26] Simon Head, *Mindless: Why Smarter Machines Are Making Dumber Humans* (New York: Basic Books, 2013).

[27] Alexander R. Galloway, *The Interface Effect* (Cambridge, UK; Malden, MA: Polity, 2012).

[28] Mark O'Connell, *To Be a Machine: Adventures Among Cyborgs, Utopians, Hackers, and the Futurists Solving the Modest Problem of Death* (Westminster: Doubleday, 2017).

[29] Sherry Turkle, *Reclaiming Conversation: The Power of Talk in a Digital Age* (New York: Penguin Books, 2015).

[30] Yuval Noah Harari, *21 Lessons for the 21st Century* (New York: Random House, 2018).

[31] Hiroshi Ishii, "Radical Atoms: Beyond Tangible Bits, Toward Transformable Materials," *MIT Media Lab*, n.d., https://tangible.media.mit.edu/project/radical-atoms/.

[32] Dr Fei-Fei Li, *The Worlds I See: Curiosity, Exploration, and Discovery at the Dawn of AI.* (New York: Flatiron Books: A Moment of Lift Book, 2023).

[33] Ajay Agrawal, Joshua Gans, and Avi Goldfarb, *Prediction Machines: The Simple Economics of Artificial Intelligence* (Boston, Massachusetts: Harvard Business Review Press, 2018).

[34] Melanie Mitchell, *Artificial Intelligence: A Guide for Thinking Humans* (New York: Farrar, Straus and Giroux, 2019).

[35] Amber Case, *Calm Technology: Principles and Patterns for Non-Intrusive Design* (Beijing Boston Farnham Sebastopol Tokyo: O'Reilly Media, 2015).

[36] Zeynep Tufekci, *Twitter and Tear Gas: The Power and Fragility of Networked Protest* (Yale University Press, 2017).

[37] Carlo Ratti and Matthew Claudel, *The City of Tomorrow: Sensors, Networks, Hackers, and the Future of Urban Life* (New Haven ; London: Yale University Press, 2016).

[38] Anthony M. Townsend, *Smart Cities: Big Data, Civic Hackers, and the Quest for a New Utopia* (New York: W. W. Norton & Company, 2013).

[39] Jeremy Bailenson, *Experience on Demand: What Virtual Reality Is, How It Works, and What It Can Do* (New York: W. W. Norton & Company, 2018).

[40] Jaron Lanier, *Dawn of the New Everything: Encounters with Reality and Virtual Reality* (New York: Henry Holt and Co., 2017).

[41] Miguel Nicolelis, *Beyond Boundaries: The New Neuroscience of Connecting Brains with Machines---and How It Will Change Our Lives* (New York: Times Books, 2011).

[42] Kate Crawford, *Atlas of AI: Power, Politics, and the Planetary Costs of Artificial Intelligence* (New Haven: Yale University Press, 2021).

[43] Rodney Brooks, *Flesh and Machines: How Robots Will Change Us* (New York: Knopf Doubleday Publishing Group, 2003).

[44] Seth Lloyd, *Programming the Universe: A Quantum Computer Scientist Takes on the Cosmos* (New York: Knopf Doubleday Publishing Group, 2007).

[45] Kai-Fu Lee and Chen Qiufan, *AI 2041: Ten Visions for Our Future* (New York: Crown Currency, 2021).

[46] Nassim Nicholas Taleb, "The Black Swan: The Impact of the Highly Improbable," *The New York Times*, April 22, 2007, sec. Books.

[47] Kevin Kelly, *The Inevitable: Understanding the 12 Technological Forces That Will Shape Our Future* (New York, New York: Viking, 2017).

[48] Mark Weiser, "The Computer for the 21st Century," *SIGMOBILE Mob. Comput. Commun. Rev.* 3, no. 3 (July 1, 1999).

[49] Walter Isaacson, *Steve Jobs* (New York: Simon & Schuster, 2011).

[50] Rajesh P. N. Rao, *Brain-Computer Interfacing: An Introduction* (New York: Cambridge University Press, 2013).

[51] Nita A. Farahany, *The Battle for Your Brain: Defending the Right to Think Freely in the Age of Neurotechnology* (New York: St. Martin's Press, 2023).

[52] Stuart Russell, *Human Compatible: Artificial Intelligence and the Problem of Control* (New York: Viking, 2019).

[53] David J. Chalmers, *Reality+: Virtual Worlds and the Problems of Philosophy* (New York, NY: W. W. Norton & Company, 2022).

[54] Hod Lipson and Melba Kurman, *Driverless: Intelligent Cars and the Road Ahead* (Cambridge: MIT Press, 2016).

[55] Erik Brynjolfsson, "The Turing Trap: The Promise & Peril of Human-Like Artificial Intelligence" (Cambridge, January 11, 2022).

[56] Hiroshi Ishii, "Tangible Bits: Towards Seamless Interfaces between People, Bits and Atoms," *ResearchGate*, n.d., https://www.researchgate.net/publication/2311166_Tangible_Bits_Towards_Seamless_Interfaces_between_People_Bits_and_Atoms.

[57] Dr Miguel A. Nicolelis and Dr Ronald M. Cicurel, *The Relativistic Brain: How It Works and Why It Cannot Be Simulated by a Turing Machine* (Natal: CreateSpace Independent Publishing Platform, 2015).

[58] Farahany, *The Battle for Your Brain*.

[59] Marcello Ienca and Roberto Andorno, "Towards New Human Rights in the Age of Neuroscience and Neurotechnology," *Life Sciences, Society and Policy* 13, no. 1 (April 26, 2017): 5, https://doi.org/10.1186/s40504-017-0050-1.

[60] Nick Bostrom and Nick Bostrom, *Superintelligence: Paths, Dangers, Strategies* (Oxford, New York: Oxford University Press, 2014).

[61] Russell, *Human Compatible.*

[62] Bruce Sterling and Lorraine Wild, *Shaping Things* (Cambridge, Mass.: The MIT Press, 2005).

[63] Brooks, *Flesh and Machines.*

[64] O'Neil, *Weapons of Math Destruction.*

[65] Harari, *21 Lessons for the 21st Century.*

[66] Noble, *Algorithms of Oppression.*

[67] Zuboff, *The Age of Surveillance Capitalism.*

[68] Zuboff.

[69] Nicholas Carr, *The Shallows: What the Internet Is Doing to Our Brains* (New York: W. W. Norton & Company, 2011).

[70] Bruce Schneier, *Data and Goliath: The Hidden Battles to Collect Your Data and Control Your World* (New York London: W. W. Norton & Company, 2016).

[71] Turkle, *Reclaiming Conversation.*

[72] Brian A. Primack et al., "Social Media Use and Perceived Social Isolation Among Young Adults in the U.S," *American Journal of Preventive Medicine* 53, no. 1 (July 2017).

[73] Naomi S. Baron, *Words Onscreen: The Fate of Reading in a Digital World* (Oxford University Press, 2015).

[74] Evgeny Morozov, *The Net Delusion: The Dark Side of Internet Freedom* (New York: PublicAffairs, 2011).

[75] Clay Shirky, *Here Comes Everybody: The Power of Organizing Without Organizations* (Penguin Press, 2008).

[76] Robert Chesney and Danielle Keats Citron, "Deep Fakes: A Looming Challenge for Privacy, Democracy, and National Security," SSRN Scholarly Paper (Rochester, NY: Social Science Research Network, July 14, 2018).

[77] Eli Pariser, *The Filter Bubble: What the Internet Is Hiding from You* (London: Penguin Press, 2012).

[78] Frank Pasquale, "The Black Box Society: The Secret Algorithms That Control Money and Information," *Book Gallery*, January 1, 2015.

[79] Aaron Perzanowski and Jason Schultz, *The End of Ownership: Personal Property in the Digital Economy* (The MIT Press, 2016).

[80] Nicholas Carr, *The Glass Cage: Automation and Us* (New York: W. W. Norton & Company, 2014).

[81] Cal Newport, *Deep Work: Rules for Focused Success in a Distracted World* (New York Boston: Grand Central Publishing, 2016).

[82] Xiao-yu Sun and Bin Ye, "The Functional Differentiation of Brain–Computer Interfaces (BCIs) and Its Ethical Implications," *Humanities and Social Sciences Communications* 10, no. 1 (November 27, 2023): 1–9.

[83] Patrick Magee, Marcello Ienca, and Nita Farahany, "Beyond Neural Data: Cognitive Biometrics and Mental Privacy," *Neuron* 112, no. 18 (September 25, 2024): 3017–28.

[84] William A. Haseltine, "Brain-Machine Interfaces Spark Ethics Debate. 4 Areas of Concern," Forbes, n.d.

[85] Marcello Ienca and Pim Haselager, "Hacking the Brain: Brain? Computer Interfacing Technology and the Ethics of Neurosecurity," *Ethics and Information Technology* 18, no. 2 (2016): 117–29.

[86] Rafael Yuste et al., "Four Ethical Priorities for Neurotechnologies and AI," *Nature* 551, no. 7679 (November 2017): 159–63.

[87] David H. Autor et al., *The Work of the Future: Building Better Jobs in an Age of Intelligent Machines* (Cambridge, Mass: The MIT Press, 2021).

[88] Zuboff, *The Age of Surveillance Capitalism*.

[89] Farahany, *The Battle for Your Brain*.

[90] Rose Luckin, Karine George, and Mutlu Cukurova, *AI for School Teachers* (Boca Raton London New York: CRC Press, 2022).

[91] Daniel J. Levitin, *Weaponized Lies: How to Think Critically in the Post-Truth Era* (New York, New York: Dutton, 2016).

[92] Sherry Turkle, *Alone Together: Why We Expect More from Technology and Less from Each Other* (New York: Basic Books, 2011).

[93] Jean M. Twenge, *iGen: Why Today's Super-Connected Kids Are Growing Up Less Rebellious, More Tolerant, Less Happy--and Completely Unprepared for Adulthood--and What That Means for the Rest of Us* (New York London Toronto Sydney New Delhi: Atria Books, 2018).

[94] Rachel Ehmke, "How Using Social Media Affects Teenagers," *Child Mind Institute* (blog), n.d.

[95] Jean M. Twenge, *iGen*.

[96] Turkle, *Reclaiming Conversation*.

[97] Sarah-Jayne Blakemore and Suparna Choudhury, "Development of the Adolescent Brain: Implications for Executive Function and Social Cognition," *Journal of Child Psychology and Psychiatry* 47, no. 3–4 (2006): 296–312, https://doi.org/10.1111/j.1469-7610.2006.01611.x.

[98] Jean M. Twenge et al., "Increases in Depressive Symptoms, Suicide-Related Outcomes, and Suicide Rates Among U.S. Adolescents After 2010 and Links to Increased New Media Screen Time," *Clinical Psychological Science* 6, no. 1 (November 2017).

[99] Stephanie S. Fredrick, Amanda B. Nickerson, and Jennifer A. Livingston, "Adolescent Social Media Use: Pitfalls and Promises in Relation to Cybervictimization, Friend Support, and Depressive Symptoms," *Journal of Youth and Adolescence* 51, no. 2 (February 2022): 361–76.

[100] David Hill et al., "Media and Young Minds," *Pediatrics* 138, no. 5 (November 1, 2016).

---

[101] Daphne Bavelier, C. Shawn Green, and Matthew W. G. Dye, "Children, Wired: For Better and for Worse," *Neuron* 67, no. 5 (September 9, 2010): 692–701.

[102] Amanda Lenhart, "What We Learned About Kids and Media in 2023 | Common Sense Media," n.d., https://www.commonsensemedia.org/kids-action/articles/what-we-learned-about-kids-and-media-in-2023.

[103] Anne-Marie Chang et al., "Evening Use of Light-Emitting eReaders Negatively Affects Sleep, Circadian Timing, and next-Morning Alertness," *Proceedings of the National Academy of Sciences* 112, no. 4 (January 27, 2015): 1232–37.

[104] Nir Eyal, *Hooked: How to Build Habit-Forming Products*, ed. Ryan Hoover (New York, New York: Portfolio, 2014).

[105] Natasha Dow Schüll, *Addiction by Design: Machine Gambling in Las Vegas* (Princeton: Princeton University Press, 2014).

[106] Michael Rich MD MPH and Teresa Barker, *The Mediatrician's Guide: A Joyful Approach to Raising Healthy, Smart, Kind Kids in a Screen-Saturated World* (Nashville, TN: Harper Horizon, 2024).

[107] Carr, *The Shallows*.

[108] Adam Alter, *Irresistible: The Rise of Addictive Technology and the Business of Keeping Us Hooked* (New York: Penguin Press, 2017).

[109] Tristan Harris, "How Technology Is Hijacking Your Mind — from a Former Insider," *Thrive Global* (blog), October 16, 2019, https://medium.com/thrive-global/how-technology-hijacks-peoples-minds-from-a-magician-and-google-s-design-ethicist-56d62ef5edf3.

[110] Sonia Livingstone et al., *Parenting for a Digital Future: How Hopes and Fears about Technology Shape Children's Lives* (Oxford, New York: Oxford University Press, 2020).

[111] Anya Kamenetz, *The Art of Screen Time: How Your Family Can Balance Digital Media and Real Life* (New York: PublicAffairs, 2018).

[112] Brandon D. L. Marshall and Scott E. Hadland, "The Immediate and Lasting Effects of Adolescent Homelessness on Suicidal Ideation and Behavior," *Journal of Adolescent Health* 51, no. 5 (November 1, 2012): 407–8.

[113] Chang et al., "Evening Use of Light-Emitting eReaders Negatively Affects Sleep, Circadian Timing, and next-Morning Alertness."

[114] Cal Newport, *Digital Minimalism: Choosing a Focused Life in a Noisy World* (New York: Portfolio, 2019).

[115] Sameer Hinduja and Justin W. Patchin, *Bullying Beyond the Schoolyard: Preventing and Responding to Cyberbullying* (Thousand Oaks, California: Corwin, 2015).

[116] Elizabeth Kandel Englander, *25 Myths About Bullying and Cberbullying / Elizabeth K. Englander.* (Wiley Blackwell, 2020).

[117] Turkle, *Reclaiming Conversation*.

[118] Claire Wardle and Hossein Derakhshan, "Information Disorder: Toward an Interdisciplinary Framework for Research and Policy Making," 2017.

[119] Jean M. Twenge, *iGen*.

[120] Susan Pinker, *The Village Effect: How Face-to-Face Contact Can Make Us Healthier, Happier, and Smarter* (New York, NY: Spiegel & Grau, 2014).

[121] William J. Doherty, *The Intentional Family: Simple Rituals to Strengthen Family Ties* (New York: William Morrow Paperbacks, 2002).

[122] Turkle, *Alone Together.*

[123] Howard Rheingold, *Net Smart: How to Thrive Online* (Cambridge, Mass.: The MIT Press, 2014).

[124] "The Future of Jobs Report 2023," World Economic Forum, accessed May 24, 2025, https://www.weforum.org/publications/the-future-of-jobs-report-2023/.

[125] Renée DiResta, "The Supply of Disinformation Will Soon Be Infinite," *The Atlantic* 20, no. 9 (2020): 2020.

[126] Madhumita Murgia, *Code Dependent: Living in the Shadow of AI* (New York, NY: Henry Holt and Co., 2024).

[127] Pariser, *The Filter Bubble*, 2012.

[128] Siva Vaidhyanathan, *Antisocial Media: How Facebook Disconnects Us and Undermines Democracy* (New York, NY: Oxford University Press, 2018).

[129] Virginia Eubanks, *Automating Inequality: How High-Tech Tools Profile, Police, and Punish the Poor* (New York, NY: St. Martin's Press, 2017).

[130] Zuboff, *The Age of Surveillance Capitalism.*

[131] Andrew Guthrie Ferguson, *The Rise of Big Data Policing: Surveillance, Race, and the Future of Law Enforcement* (New York: NYU Press, 2017).

[132] Brynjolfsson and McAfee, *The Second Machine Age.*

[133] Carl Benedikt Frey and Michael A. Osborne, "The Future of Employment: How Susceptible Are Jobs to Computerisation?," *Technological Forecasting and Social Change* 114 (January 1, 2017): 254–80, https://doi.org/10.1016/j.techfore.2016.08.019.

[134] Frey and Osborne.

[135] Agrawal, Gans, and Goldfarb, *Prediction Machines.*

[136] Lee, *AI Superpowers.*

[137] Richard A. Clarke and Robert K. Knake, *The Fifth Domain: Defending Our Country, Our Companies, and Ourselves in the Age of Cyber Threats* (New York: Penguin Press, 2020).

[138] Paul Scharre, *Army of None: Autonomous Weapons and the Future of War* (Erscheinungsort nicht ermittelbar: W. W. Norton & Company, 2018).

[139] Klaus Schwab, "The Fourth Industrial Revolution."

[140] Klaus Schwab, Nicholas Davis, and Satya Nadella, *Shaping the Future of the Fourth Industrial Revolution* (New York: Crown Currency, 2018).

[141] "A New Chapter in E-Estonia's Development," TI-Hüpe, n.d., https://tihupe.ee/en.

[142] OECD, *PISA 21st-Century Readers Developing Literacy Skills in a Digital World* (Paris: OECD, 2021).

[143] Russell, *Human Compatible.*

[144] Tegmark, *Life 3.0.*

[145] Vincent Boulanin and Moa Peldán Carlsson, "Limits on Autonomy in Weapon Systems: Identifying Practical Elements of Human Control" (Stockholm: Stockholm International Peace Research Institute, 2021).

[146] Russell, *Human Compatible*.

[147] Virginia Dignum, *Responsible Artificial Intelligence: How to Develop and Use AI in a Responsible Way* (Springer Nature, 2019).